About Island Press

Since 1984, the nonprofit organization Island Press has been stimulating, shaping, and communicating ideas that are essential for solving environmental problems worldwide. With more than 1,000 titles in print and some 30 new releases each year, we are the nation's leading publisher on environmental issues. We identify innovative thinkers and emerging trends in the environmental field. We work with world-renowned experts and authors to develop cross-disciplinary solutions to environmental challenges.

Island Press designs and executes educational campaigns in conjunction with our authors to communicate their critical messages in print, in person, and online using the latest technologies, innovative programs, and the media. Our goal is to reach targeted audiences—scientists, policymakers, environmental advocates, urban planners, the media, and concerned citizens—with information that can be used to create the framework for long-term ecological health and human well-being.

Island Press gratefully acknowledges major support of our work by The Agua Fund, The Andrew W. Mellon Foundation, The Bobolink Foundation, The Curtis and Edith Munson Foundation, Forrest C. and Frances H. Lattner Foundation, The JPB Foundation, The Kresge Foundation, The Oram Foundation, Inc., The Overbrook Foundation, The S.D. Bechtel, Jr. Foundation, The Summit Charitable Foundation, Inc., and many other generous supporters.

The opinions expressed in this book are those of the author(s) and do not necessarily reflect the views of our supporters.

What Should a Clever Moose Eat?

What Should a Clever Moose Eat?

NATURAL HISTORY, ECOLOGY, AND THE NORTH WOODS

John Pastor

Illustrated by the author

Washington | Covelo | London

Island Press is a trademark of The Center for Resource Economics.

Library of Congress Control Number: 2015937835

⚜ Printed on recycled, acid-free paper

Manufactured in the United States of America
10 9 8 7 6 5 4 3 2 1

Keywords: Adirondacks, balsam fir, beaver, blueberries, boreal forest, fire, food web, glaciation, ice sheet, Lake Superior, life cycles, life histories, Maine, maple, Minnesota, moose, moraines, natural history, North Woods, pine, pollen analysis, spruce budworms, spruce, tent caterpillar, voyageurs

For my grandson,
Laszlo Pastor

Everything changes; everything is connected; pay attention.
   —Jane Hirshfield

Contents

List of Drawings

Foreword

Bernd Heinrich

Eighteen thousand years ago most of northern North America was covered by the ice of mile-deep glaciers. By ten thousand years ago the ice had melted, creating a tundra habitat with ponds, lakes, eskers, moraines, and drumlins. A mere six thousand years ago this tundra became the North Woods, an ecosystem populated by beaver, hare, lynx, and moose, encompassing New England, Quebec, Ontario, Labrador, Wisconsin, and northern Minnesota.

For the past 30 years, John Pastor has studied the North Woods, and his book *What Should a Clever Moose Eat?* showcases his deep knowledge of this region. His insightful essays explore natural history questions and observations and delve deeply into how this ecosystem works. By extension, this book is not only about the North Woods; it is about the larger theme of how natural systems work and how we go about finding good questions and the answers to them.

The North Woods are "simple" in having fewer but often prominent species, allowing us to more easily observe how species affect the structure and complexity of the ecosystem. What a moose eats has easily measurable consequences, and as the chapter on moose illustrates, a single moose has a large effect on the cycling of nutrients through an ecosystem. Moose thus bequeath not only their genes to the next generation but also the landscape they created. The theme comes through in several other iconic and famous examples throughout this book, as we see this ecosystem in terms of its origin, its development, and its parts today.

Several stories in this book reveal effects that could not have been predicted. For example, conifers apparently do not invade meadows created by beavers. From one natural history observation and one experiment that led to the next, scientists determined that mycorrhizae and red-backed voles are critical variables, and together they answer the riddle of the excluded conifers. Similarly, studies revealed that tent caterpillars' periodic mass outbreaks are the result of arms races in which broad-leaved trees defend themselves against caterpillars, luring caterpillar-killing ants with sugary secretions. On a longer time scale, the population explosions and crashes of both hares and lynx are cyclical, and they have become a model system for studying populations thought to be related strictly through food. Exhaustive experiments indicated a correlation with sunspots.

This long time scale often hinders our understanding of an ecosystem. Predator–prey cycling between warblers and spruce budworm affect native northern spruce–fir forests over the course of a half century. But budworm defoliations could be made nearly chronic by new methods of timber harvesting and, counterintuitively, by the use of biocides to kill the budworms. As Pastor's examples show, tinkering with the ecosystem leads to surprises because of its unanticipated complexity. What applies to control of biological agents in a natural ecosystem applies to fire as well, and Pastor devotes a few chapters to showing how fire and browsing animals have left their imprint on the North Woods.

At least since the days of the fur trade, humans have left a vast imprint on the region. They began by nearly eliminating the beavers, but they also opened the region to explorers who returned not only with pelts but with new observations and discoveries that had consequences for the North Woods and beyond. The Maine woods fascinated Henry David Thoreau, who in turn inspired George Perkins Marsh of Vermont, whose famous book *Man and Nature* provided a warning about the human imprint and scientific rationale for conservation, which led to the designation of the Adirondack Preserve in 1892 to preserve its lands as "forever wild." The Adirondack model has been applied worldwide since then, as we realize that we simply do not and will not ever be able to understand it all. The North Woods provided a world model, and does so still.

"What should a clever moose eat?" is not just a rhetorical question. The answer has large consequences not only for individual moose but also for moosedom and entire ecosystems. Answering involves ecological, physiological, and evolutionary studies. But these studies would be futile, and even wrong, if not based on good natural history questions sparked by careful observations in the field. If nature teaches us anything, it is that it is complex, and seemingly little things can have huge consequences.

Preface

During the past several decades of my research on the forests, peat-lands, beaver, moose, and other occupants of the North Woods, the idea slowly dawned on me that every good problem in ecology begins with a question or observation from natural history. To explore this idea, I began writing occasional essays about how organisms that live in the North Woods teach us something larger about the ecology of the world around us. These essays were short, about a thousand words, and were published many years ago in newsletters of the Voyageurs National Park Association and the Friends of the Boundary Waters Wilderness and in *Minnesota Forests*. In reading through these old essays a few years ago, I decided that I could rewrite and expand them to bring them up to date. As I did this, it occurred to me that if I wrote some more essays, I would have a book on the natural history of the North Woods and why and how natural history underlies all of ecology and most of biology and geology.

The result is this set of essays about the natural history of the North Woods and how natural history helps us understand the world better. An essay (from the French word *essayer*, "to try" or "to attempt") is a foray into the unknown, an attempt to find some order in one's world. This attempt to find order in the unknown is exactly the spirit with which natural history is done. In this book I want to explore the varieties of approaches to nature that constitute good natural history thinking and where these approaches lead us, using the North Woods as my arena. In a way, the entire book is one extended attempt, or essay, to make sense of what we mean by "natural history" and why I and many others have come to appreciate anew its importance to the fields of ecology in particular and biology and geology in general.

This book takes a "layer cake" approach to the natural history of the North Woods. I begin at the lowest level of the cake with the formation of the landscape by the ice sheet that covered most of the current location of the North Woods 10,000 years ago. The ice sheet left a variety of landforms that determine the distribution of water, especially water held in the soil. The water in the soil in turn determined how the plants of North Woods, the next layer of the cake, assembled themselves into communities as the ice sheet retreated and they migrated into the land that emerged from beneath the ice sheet. Soon after, herbivores, pollinators, and seed dispersers arrived and formed the third layer of the cake. The fourth layer, the predators, arrived at about the same time as or immediately after the herbivores and pollinators. Native Americans and then European trappers and fur traders arrived, followed by loggers, and we became a fifth layer. Fires, especially in the drier part of the North Woods in the western Great Lakes region, periodically reset the landscape to bare soil, and the North Woods reassembles itself anew.

The main point of this book is that the North Woods or any other ecosystem is not simply a collection of species or a series of layers atop one another. The point I wish to make is that the connections between these species and layers—how water, energy, and nutrients flow between them, who eats whom and why, how different plant species promote or impede fires, how evolution continually shapes and reshapes these connections—are determined by the natural history of the organisms. Natural history is more than just identifying species and knowing what habitats they are found in, although these are excellent starting points. The natural history of an organism includes its life cycle, how it responds to the seasons, what it eats and who eats it, how it reproduces, where it gets its energy and nutrients from, where the energy and nutrients go when they die, and how these traits have coevolved with other organisms in the food webs and ecosystems to which they belong.

The book takes you, the reader, up through the layers of this cake. Each set of essays builds on and elaborates topics described in previous essays. All the essays begin with some observation or question in natural history; those of us who live in the North Woods will be familiar with many of these just from our walks through the woods. After two intro-ductory essays on the nature of natural history and the North Woods, Part I describes how the communities of plants and animals of the North Woods assembled themselves on the watery landscape the ice sheet left behind and how one important animal, the beaver, modified and still controls the flow of water. It closes with an essay on how one brilliant observer in the fur trade laid the foundation for the study of the nat-ural history of the North Woods. Part II discusses how the life history and shapes of leaves and crowns of different plant species captures the sun's energy. Part III shows how the natural histories of herbivores and predators take the energy captured by plants and weave it into the food web. Part IV discusses how plants reproduce themselves through flowers and seeds, how animals pollinate, eat, and disperse these seeds, and how these reciprocal relations between plants and animals have evolved. Part V explores the idea that fire is a process that renews rather than destroys the North Woods and that some species have become adapted to and even need periodic fires to release and disperse seeds. A common thread throughout these essays is how natural selection has knitted these lay-ers and connections between species into a functioning ecosystem. The book closes with an epilogue on the currently changing climate and the future (or demise?) of the North Woods and with a postscript of afterthoughts on how we humans evolved an interest in the natural history of the world around us.

I hope this book finds a wide audience that includes intelligent non-scientists interested in the natural world, undergraduate and graduate students looking for potential research problems, and my colleagues. Although this is not a textbook about the North Woods, it could be used as supplementary text in ecology classes or as the main text in seminar classes. It could also be a sourcebook of ideas for research projects. I have tried to walk a fine line between scientific rigor and losing readers who may not be familiar with some scientific terms and concepts. Scientific terms are defined explicitly, by example, or by both when first mentioned, and there is a short glossary of **bold-faced** words at the end of the book. I have used metric units throughout rather than English because that is the accepted practice of science. For those who are not familiar with metric units, here's a brief introduction: A millimeter is the width of a spruce needle, the length of the fingernail on your pinkie is about a centimeter (if you keep it trimmed), a meter is about 3 inches longer than a yard, a kilometer is 6/10 of a mile, a hectare is about 2½ acres, a liter is slightly more than a quart, water boils at 100°C and freezes at 0°C, and −40°C is the same temperature as −40°F.

On the other hand, except in a few places, I have not used the scientific Latin names for plants and animals, such as *Betula papyrifera*, as is accepted taxonomic practice. The reason why I have generally avoided Latin names is that they would unnecessarily lengthen lists of species when I give them and also clutter the text. Beginning with Linnaeus in the 1700s, scientists began using Latin rather than common names because there were at that time so many common and colloquial names for many species, such as *white birch, paper birch, canoe birch,* or *silver birch* for *Betula papyrifera*, the familiar white-barked species depicted on the cover. The Latin names also have the advantage of indicating evolutionary relationships between species because the first name (*Bet-*

ula) indicates the genus within which closely related species depicted by the second name (*papyrifera* along with its relatives *alleghaniensis, cordifolia, nigra,* and *pumila*) are grouped. But with the development of field guides such as the Peterson series, common names have become less colloquial and more standardized, so *paper birch* is the accepted common name now for *Betula papyrifera*. If you want to learn the Latin names and more about the natural history of any species mentioned in this book, you can easily find them by entering the common name I use into a web browser. I recommend starting with *Wikipedia* and following up with the references at the end of the species page. The few exceptions where I use Latin names are where there are no standard common names for the species being discussed.

Scientists have learned to speak about the natural world using words such as *perhaps, maybe,* and *possibly* rather than in definite and absolute terms. This can frustrate nonscientists; there have been many times when people have asked me, "Why can't you scientists just give us the answer?" The best scientists are by nature cautious. We are reluctant to make any firm pronouncement because we know it is always possible that a few more observations could force us to change our minds, or at least to rethink something. Instead we offer hypotheses, which are tentative explanations subject to change or abandonment when confronted with data, but starting sentences with "I hypothesize that…" sounds stilted. Using words such as *perhaps, maybe,* and *possibly* seems to me to be a gentler way to invite you to consider and explore a different way to view the world.

The scientific studies that supplied particular experiments or observations mentioned in the text are referenced in Notes at the end of the book. These Notes give the author's last name and date of publication. You can find the full references in the Bibliography at the end of the book, which is organized according to the title of each essay. Many of these publications can be found online using search engines such as

Google Scholar, although you may need to be logged in through a university library to obtain the full text. These references should help any student or colleague who wants to follow up on some of these ideas. If I have mistakenly explained any research mentioned in these essays or provided the wrong citation, please let me know.

I could not have written this book without the advice and support of several organizations and people. Almost all of my research and that of many of my colleagues mentioned in this book has been funded by the National Science Foundation. This is one of the very best-run government agencies; more than 90 percent of the money it receives from Congress goes out to support students, technicians, faculty, and other scientists in their quest to understand the natural world. The National Science Foundation is an agency run by scientists for scientists. It is of great credit to this country that the National Science Foundation was established more than 50 years ago. Every new administration claims that they want to double funding for research supported by the National Science Foundation. I hope someday that actually happens.

My time writing this book was supported by a generous sabbatical leave from the Swenson College of Science and Engineering of the University of Minnesota Duluth. I thank Dean Emeritus James Riehl for awarding me this sabbatical and for his support of my work over the years.

Two people have read through every essay and have saved me from much grammatical embarrassment but more importantly called my attention to places that needed clarification so that you, the reader, would not be lost. They each deserve more thanks than I can express. The first is my wife, Mary Dragich, who corrected mismatches between verb tense and subject (there were many) and helped me clarify my thoughts. Mary also suggested several excellent examples that helped illustrate the points I was trying to make. I am ever grateful for her patience with me while I was working on this book and her patience and love at all other times as well.

Barbara Dean, my editor at Island Press, also read through every essay and made many helpful suggestions for clarification of difficult passages. Most importantly, Barbara helped me shape this book from a loose collection of essays with a staple through them (my first approach) to what I hope is a more coherent book that still maintains the informal and exploratory quality of a set of essays.

I also thank other members of Island Press for their unfailing assistance and support for this book. These include most notably Erin Johnson and Rebecca Bright for production support, David Miller, the publisher, and Maureen Gately, who designed the beautiful cover. It has been a pleasure working with all of you.

PROLOGUE:
The Beauty of Natural History

Why every good question in ecology and geology begins with an observation and question in natural history.

Naturalists and ecologists often ask questions that most people consider, well, peculiar. Such as "What should a clever moose eat?" When I once told my sister, who is not a scientist, that this was the question I was working on at the time, she looked at me and said, "Anything it wants to. Why should you care?" My sister asked a good question. The purpose of this book is to explain why she and other people who aren't scientists, as well as other scientists, should care about the questions naturalists and ecologists ask.

Natural history is about particular organisms living in particular places, and the place of this book is the North Woods, the mixture of boreal conifers and northeastern deciduous species that stretches from Minnesota to Newfoundland. The North Woods is one of the most ecologically, geologically, and aesthetically interesting places on Earth. Here, glacial deposits from the Ice Ages lie atop some of the oldest rocks on Earth and support northern hardwoods and boreal conifers. Here, we find moose and beaver, the highest diversity of birds north of Mexico, and species with spectacular population dynamics such as spruce budworm and balsam fir, lynx and hare, wolves and moose, and many others. These properties of the North Woods give us room to explore the breadth and depth of natural history and how it helps us frame ecological research.

The importance of natural history in initiating and guiding research questions is a recurrent theme throughout these essays. I think it was Dan Janzen who once said (I can't find where) that everything hinges on the details of natural history. The common threads through these essays include how the physical environment constrains ecological processes, how species interact with each other and the landscape, and how **evolution** by natural selection modifies these interactions.

Natural history asks questions such as "What does this organism do?" "What is its life cycle?" "How does that help it survive and reproduce?" "What does that mean for the other organisms it interacts with and the landscapes it inhabits?" "How does this landscape work, and how did it come to be?" Answering these questions requires sustained efforts to describe organisms and the places in which they live. The sustained collection of data to build a rich description of the patterns of a place was once derided by James Watson as "stamp collecting," but out of it can come rich and deep theories. Darwin did not set out to construct a theory of evolution in which sex and variation played prominent roles. Instead, the importance of sex in maintaining variation came to him during the course of describing all known living and fossil barnacles and their life histories.[1] The theory of natural selection and evolution then emerged and grew from Darwin's lifelong work on barnacles, pigeons, orchids, worms, and other species.

During the past 30 years in northern Minnesota, I have spent a lot of time thinking, "Why does a moose eat this plant and not that one, and what difference does it make?" The papers that my students, colleagues, and I have written are attempts (essays?) to answer them. For many years, we collected data to see what happens to the plant community and the soils as moose populations wax and wane. We also used GPS collars, autoanalyzers, and gas chromatographs; developed mathematical models of moose energetics, plant growth, and nutrient cycling; analyzed data with a wide variety of powerful statistical techniques; and developed several competing hypotheses about what might happen. What often did happen was entirely different from any of these hypotheses. Nature is always more interesting than the hypotheses we first propose. Natural history teaches us to hold our theories and hypotheses lightly, to keep open to new observations, to not anticipate too rigidly what we are supposed to observe or measure. My questions—why does a moose eat this plant and not that one, and what difference it makes—emerged from watching and reading and thinking about what a moose does. In short, the natural history of moose suggested the questions and our experimental and mathematical approaches to answering them.

It is my firm belief that any good research problem in ecology and probably most of biology and geology begins with a natural history question or observation. There are many ways to do science, not just The Scientific Method of hypothesis testing, which we teach in freshman classes, with capital letters in the tone of our voices. Hypothesis testing is a powerful way to do science, to be sure. But before we can offer a hypothesis, design an experiment, or construct a model, we need to have a clear description of what it is about nature that we are trying to understand. Any hypothesis developed in a vacuum devoid of clear observations and descriptions will not be very interesting. When done at its best, natural history provides descriptions of nature that cry out for explanations and answers to some interesting observation, to some anomaly or asymmetry, or to some surprising questions about how an organism or landscape works. For instance, our understanding of how the North Woods arrived in its present location came from patiently constructed descriptions of pollen sequences in cores from hundreds of lakes over many years, one lake and one core at a time. What surprised us was that the range of each species expanded northward independently and sometimes even at right angles to the expansions of others. In the beginning, no one would have come up with range expansions in perpendicular directions as a hypothesis of species migrations, but that's what happened. It still cries out for an explanation. As Walter Tschinkel and E. O. Wilson remarked, natural history produces "an abundance of serendipity" that keeps our minds fresh.[2]

Once a hypothesis is developed, we examine it further by experimental manipulation of one or more factors. Or we develop a mathematical model to work through whether the hypothesis logically follows from its premises. Or we use the model to make predictions that can be experimentally tested. But even in the experimental or modeling phase of the research, natural history is always guiding the decisions we must make. What portion of an organism's life cycle should be the focus of our experiments? What factors should we manipulate, and what levels of experimental treatments should we impose on the organism to test the hypothesis? What are the biological meanings of the equations in our mathematical models? There are an infinite (or at least very large) number of answers to these questions, but we make our choices guided by what we know about the natural history of the organism. For example, there is little point to testing the responses of an organism to nutrient concentrations well outside the range that the organism experiences in the wild. In order for a mathematical model to sharpen our biological insight, every variable and parameter must have a biological meaning that is almost always a property of the natural history of an organism, such as birth rate and lifespan, among many others. Natural history thinking is not done simply to describe nature or generate questions; it is (or should be) present throughout the course of research. I have confidence in scientific conclusions when natural history observations, experiments, and mathematical models all lead to the same conclusions, but especially when natural history is the thread that binds them together. Always, we return to natural history.

Natural history is often defined as the study of organisms and their ecology,[3] but organisms live in landscapes (or seascapes), and these have a geological history. Natural history bridges observations of an organism's life cycle, behavior, and interactions with other organisms with the geological history and structure of the landscapes in which it lives. Darwin was both a geologist and biologist, in modern terms, but in his letters, especially later in his life, he most frequently characterized himself as a natural historian.[4] Darwin was the Grand Master at synthesizing keen observations on the natural history of organisms, mountain ranges, and coral reefs into powerful theories. *The Voyage of the Beagle* is best known for the chapter on the Galápagos. But the chapters on the rise of the entire Andean Cordillera are the ones that best demonstrate how Darwin developed general theories from natural history observations. In these chapters, Darwin showed how today's Andes could be explained by the many small repeated uplifts of the land in earthquakes similar to the one he experienced in Valparaíso, Chile. He later used the same type of reasoning to develop a theory of how the natural history of individual coral polyps controls the formation of massive coral reefs and atolls. The common thread through these studies was that large effects (mountain ranges, coral reefs) result from the accumulation of small repeated changes (earthquakes, growth and reproduction of individual coral polyps) over long periods of time. This thread later became a linchpin of Darwin's theory of evolution in the *Origin of Species*, which is one long catalog of natural history observations and questions that led not only to his theory of natural selection but also to several other general theories about how nature is organized. If you haven't browsed through the *Origin* recently, take a look at it. You will be pleasantly surprised. My favorite passage is when Darwin, after musing on red clover and violets, humblebees, mice, and cats, invents the concept of a **trophic cascade**.[5] Much of our research today is a series of footnotes to Darwin's natural history theories.

Natural history questions often arise from simple, serendipitous observations that anyone can make on a walk through the woods. Sometimes when a student comes to me asking for a research problem, I suggest that he or she take a walk in the woods and find a plant that intrigues him or her, even if only because it has pretty flowers. The aesthetic beauty of an organism will at least keep the student's attention focused until he or she develops a sense of the beauty of the question he or she will try to answer. I then give the student a List of Questions to Ask a Plant, which includes "Who pollinates you?" "How do you grow in full sun and in shade?" "How do you grow on different soils?" "How many seeds do you make?" "Do you make the same number of seeds each year?" "Who disperses your seeds, and when and where?" "Who eats you?" "Why do they eat you?" "What other plants do you associate with?" The answers to these questions, and more, describe the natural history of the plant. Unfortunately, we do not know these answers for most plants. Our predictions of how nature will respond to timber harvesting, climate change, hunting and gathering, invasion by exotic species, and other factors are severely limited by not having answers to these questions.

Although serendipity may put you in the right place at the right time to make a natural history observation, it does not guarantee you will ask a good question because of it. You have to be able to recognize that there is a good question lurking in the observation. What makes a good natural history question? This is harder to answer than it may at first appear. To a great degree, the value of a natural history question lies in the surprises generated by the research that attempts to answer it, so sometimes we only find out if it is good long after we ask it.

Many good natural history questions arise from observations made at human scales, such as the behavior of whole organisms or the landscapes they reside in (well, maybe the human needs a microscope to observe zooplankton in a lake, but you get the idea). We widen the scope of these observations when we use them as springboards to make inferences or pose new questions about how the world works. This is the point where surprises often arise, and before we know it, we are going in a direction we didn't originally think we would travel. These new directions almost always connect our original observation with other levels in the biological hierarchy. Sometimes a natural history observation leads us to think about molecules, and sometimes it leads us to contemplate how the entire globe works. But the fascination of a natural history question is that *we do not know where and in what direction it will take us when we first ask it.* If we ask a molecular question, we are likely to get a molecular answer. This is not to disparage molecular biology, but it is to say that because natural history questions start in the middle of the range of the scales of nature, it allows us to go in more directions than if we started at one or the other extreme. Furthermore, we can keep coming back to our original natural history question, approaching it from several angles or from smaller or larger scales, and connecting it with other related questions. Anyone could spend a lifetime doing this with just one question, if it is a good one (see especially John Bonner's scientific memoir of his fascination with slime molds[6]). By focusing on questions at the organismal and human scale, natural history can help integrate the many different levels of biological organization, which are difficult to bridge because of different techniques, goals, and criteria for what constitutes excellence in research.[7]

Not every question about the natural world is a good natural history question. "How many leaves are on this tree?" is not an interesting natural history question because, although you can certainly answer it, it doesn't lead to other levels of biological organization. In short, it gets you nowhere. In contrast, "Why are leaves dispersed throughout an aspen's canopy but arranged in a shell at the perimeter of a maple's canopy?" is a very good natural history question. Although anyone can make this observation, it was Henry Horn who developed it into a theory of the adaptive geometry of trees.[8]

A reason why this second question intrigues us (and perhaps intrigued Henry Horn) is because of the asymmetry in the different arrangements of leaves in aspen and maple: Why are the leaves arranged in opposite ways in these two species, which often grow together? Horace Freeland Judson called such asymmetric properties of nature that capture our attention "broken symmetries."[9] Broken symmetries catch our eyes, cause intellectual dissonance, and compel us to resolve them. Good natural history questions are often based on a broken symmetry that leads us in many directions. The broken symmetry between aspen and maple leaf arrangements also makes us think of other asymmetries between them, such as their very different shade tolerances and photosynthetic responses to light, the ovoid shape of an aspen leaf compared with the lobed shape of a maple, the differences in their relative growth rates, and many other properties. What are the connections between all these broken symmetries between aspen and maple? How did these evolve? How do they explain how aspen and maple adapt to their environments? Do they explain why maple succeeds aspen as a forest recovers from clearcutting? Do they determine aspen's and maple's very different roles in the food web and in the ecosystem? If you can't sleep at night until you understand what a broken symmetry is saying, you know you have a very good natural history question by the tail. The question that has kept me awake many nights—why does a moose eat this and not that?—contains a very powerful asymmetry whose answers were not obvious when we first asked it and even today are not completely known. Seeking those answers has led us in many directions, such as investigating plant and soil chemistry, the physiology of moose energetics, how plants respond to being eaten, and the development of spatial patterns of browsing and plant distributions across the landscape. But I keep coming back to it from new angles, each initiated by thinking about a different aspect of the natural history of moose. Right now, I am thinking about how the changes to the forest ecosystem wrought by moose browsing become selection pressures on future generations of moose.

Natural history is also a good way for children and nonscientists to enter into the scientific study of nature. It is not necessary that nonscientists and children engage in elaborate experiments or mathematical models or that they become coauthors on scientific publications. But learning how to pose a good natural history question from observations made during a walk in the woods or a day weeding the garden would add an extra dimension and enjoyment to these activities. It would also enrich children's and citizens' understanding of nature and how we do science. We desperately need informed citizens who understand nature and the practice of science, because making sound environmental policy requires that citizens who vote and policymakers who write laws have some idea of how nature works and how we find out about it. Nonetheless, the place of natural history in school and college curricula has declined drastically in recent decades.[10]

As we become more urban, many people, especially children,[11] are becoming increasingly estranged from nature. Yet natural history underlies many of today's policy and legislative issues, including global warming, the sustainable harvesting of resources, the control of predators and insects, and the preservation of species. Natural history is the underpinning to conservation,[12] to natural resource management,[13] and to human health and food supply.[14] We have learned precious little about the natural history of most organisms other than those we can harvest for money, even in biomes as well studied as the North Woods. Much, if not most, of what we know about the nonharvestable flora and nongame wildlife of any biome is the paragraph on the organism in a field guide. But how the North Woods or any other biome will respond to climate change, timber harvesting, hunting, or invasion by exotic species depends on the details of the natural history of its constituent organisms. We need to help people re-engage a sense of delight and wonder in the natural world to address these practical problems. The natural history of where they live is probably the best place to start.

A fascination with how the world works is an important part of what it means to be a human being; the cave paintings at Chauvet testify eloquently to that.[15] If we lose that fascination, we become less human. Our decisions on how we care for the earth will be enhanced if we renew this basic human trait. As Robert Michael Pyle has said, "What we know we may choose to care for. What we fail to recognize, we certainly won't."[16] I firmly believe that if people could know, really know, what beautiful, living, working systems lakes, rivers, prairies, wetlands, beaches, and forests are, they would do everything they could to preserve them. Causing the extinction of a species or the demise of an ecosystem would then seem a crime equal to the defacing of the *Mona Lisa* or the *Pietá*. I hope this book helps people understand that.

INTRODUCTION
The Nature of the North Woods

What we can learn from the natural history of the North Woods.

I am standing on the western shore of Lake Superior in Duluth, Minnesota and looking east. I am not far from where the French fur traders known as the voyageurs beached their freighter canoes after paddling them 2,000 kilometers up the chain of Great Lakes from Montreal. Pause now to look at a map of the western Great Lakes region, or go vicariously to the western shore of Lake Superior near Duluth using Google Earth. Ahead of me, across the western arm of Lake Superior, is the Bayfield Peninsula of Wisconsin. East of that, for 3,000 kilometers, the same length as the Amazon Basin, the North Woods stretches continuously to the far coast of Newfoundland, the easternmost point of the North American continent. Two hundred and sixty kilometers west of me, the North Woods meets its abrupt boundary with the prairie. This is an immense forest, spanning an area of approximately 2 million square kilometers.

The North Woods is one of the most ecologically, geologically, and aesthetically interesting places anywhere. Here, the geologically youngest glacial deposits from the Ice Ages lie atop the Canadian

1

Shield, which contains some of Earth's oldest rocks. The North Woods contains the southern portions of the ranges of boreal tree species such as white, black, and red spruce, balsam fir, tamarack, paper birch, and quaking aspen. These overlap with the northern portions of the ranges of white, red, and jack pine, hemlock, white cedar, red and sugar maple, yellow birch, northern red oak, basswood, beech, black cherry, white and green ash, mountain ash, and other deciduous species. It is also a forest with an abundance of fruit-bearing shrubs, small trees, and vines such as blueberries, lingonberries, raspberries, juneberries, wild plums, cranberries, and wild strawberries. More simply but perhaps less precisely, the North Woods is the band of forest centered on the Great Lakes and the St. Lawrence River where the range of sugar maple to the south overlaps the range of balsam fir to the north. This is the land of Christmas trees and maple syrup.

It is sometimes said that the North Woods is a transitional forest between the boreal forest and the eastern deciduous forest. Although this may be technically true, it does not capture the integrity and complexity of the North Woods as an intact ecosystem in its own right. It is like saying Beethoven was a transitional composer between Mozart and Schubert. Except for smaller mixed forests of northern coniferous and deciduous tree species in Scandinavia, the French Alps, northeastern China, and Hokkaido in Japan, there is no other forest anywhere on Earth like the North American North Woods. As you drive north from Minneapolis, Chicago, Detroit, Toronto, New York, or Boston, you know the exact point on the road where you have entered the North Woods. The character of the forest suddenly changes wherever the more southerly deciduous species meet the more northerly conifers. There is a clear sense of arrival when the tang in the air, the dark wall of conifers, perhaps the sight of an eagle or a beaver pond, all unite to proclaim *North*. Few, if any, people ever think "Now I am entering a transitional forest."

The Seasons of the North Woods

The North Woods is a very beautiful forest, especially in autumn when the luminous reds and yellows of deciduous trees contrast with the dark blue-greens of the conifers, but each season has its own beauty. Living in the North Woods through its spectacular changes across the four seasons can feel like living in four different ecosystems in succession each year.

The temperature of the North Woods ranges from as cold as $-40°C$ in winter to as high as $35°C$ in summer. The North Woods in winter is colder than anywhere else on Earth at the same latitude, even much colder than northern Scandinavia, which is 15 or 20 degrees farther north. In winter, low-pressure systems from the northern Rocky Mountains or the Alberta plains sweep across the Dakota prairies and then Minnesota, the Great Lakes, Ontario, Quebec, the Adirondacks, New England, and the Maritime Provinces. The winds in these low-pressure systems circulate counterclockwise. As they pass by, the north winds on their back sides draw polar air masses down from the Arctic, there being no high mountain ranges north of the Great Lakes to block their flow. If the low is followed by an Arctic high-pressure system circulating clockwise, then the north winds at the front of the high augment the northerly winds on the back of the low, bring a sustained flow of polar air into the North Woods. This is when the coldest temperatures happen.

The average southernmost extent of this very cold polar air mass in winter, a line from approximately the middle of Minnesota through New England and the Maritime Provinces, defines the southernmost boundary of the North Woods. Six months later, as storms sweep up from the southern plains and Gulf of Mexico, polar air masses are pushed northward. The rotation of Earth usually bends the trajectories of these southerly air masses eastward before they reach well into the North Woods. Summers in the North Woods remain generally pleasant, having only a few days or a week of the very hot and humid weather that plagues the rest of North America east of the Mississippi. This is why tourists like to come to the North Woods in summer: to get north of those humid air masses.

The seasonal changes in these weather patterns are accompanied by large changes in day length from fewer than 8 hours at the winter solstice to more than 18 hours at the summer solstice. In the North Woods, daylight is as much a seasonal as it is a daily phenomenon. The wide swings in weather and day length over the course of the year are cues for organisms to complete milestones in their annual cycles of development and reproduction. **Phenology**, or the compiling of long-term observations of the timing of these milestones in a species' life cycle, is an important aspect of the study of the natural history of the North Woods. By comparing today's phenological observations with those recorded by Thoreau, for example, we have learned that climate warming in the past century and a half has advanced the time of first flowering in the North Woods by at least 10 days and even 2 weeks in some cases.[1] The phenology of development and reproduction during a species' life cycle has evolved through natural selection to be in concert with the phenology of other species with which it interacts. Pollinators and flowers, for example, need to have their developments synchronized for both to benefit. However, we know little about what possible phenological changes in timing of flowering, initiation and cessation of growth, and migration induced by climate change will mean for the evolution of local populations and the productivity and cycling of nutrients in the entire ecosystem. Asynchronous shifts in the phenology of plant flowering and the appearance or arrival of their pollinators could spell doom for local populations of both. The evolutionary consequences of climate change may lie in these disrupted phenologies between plants and pollinators.

On the spring equinox, the North Woods is usually still encased in winter's full accumulation of snow, although the −20 to −40°C temperatures are probably over. As I write this on the spring equinox of 2014, we are having a steady snowstorm that is forecast to dump at least 30 centimeters on top of the meter or more already on the ground. Tomorrow, the temperatures will plummet to –20°C as the low passes by and polar air is dragged southward. Although the chickadees began their "fee bee, fee bee" mating calls in February, those calls were more of a down payment on spring than the actual start.

Spring starts when we can smell the soil. Actually, what we smell is not the mineral soil or humus but volatile compounds called geosmins being released by actinomycetes. Actinomycetes are bacteria that produce fine rootlike threads called mycelia that help them decompose organic matter and gather nutrients from it. The smells of the geosims emitted by actinomycetes are signs that the decomposition of last autumn's leaf fall is beginning to jumpstart the cycles of nutrients that will fuel plant production over the coming growing season.

At the same time, maple sap is flowing and leaves are emerging from leaf buds. Although the autumn colors are spectacular, spring can be equally colorful in its own way. Deciduous leaves of a startling variety of pastel greens are emerging from twigs whose reddish browns also seem to brighten. When these emerging pastel greens are set off by the cinnabar reds of maple flowers, the darker greens of conifers, and the sprays of white flowers on branches of juneberries and wild plums, a forested hillside can be truly stunning. During the few weeks when the emerging leaves are no larger than a squirrel's ear, nearly full sunlight penetrates the canopy to the forest floor. Stimulated by the light and warmth, the spring flora rapidly complete their short flowering periods. The forest floor is awash with white and pale pastel flowers of hepaticas, bloodroots, trilliums, bunchberries, anemones, spring beauties, merrybells, violets, true and false Solomon's seals, twinflowers, and ladyslippers, moccasin flowers, and other orchids, among many others.

As if in competition with the floral display, the males of the returning two dozen or so species of warblers are flashing a kaleidoscope of primary colors as they establish territories, build nests, and court females. Eagles, ospreys, geese, ducks, swans, and herons arrive and search for ice-free water where they can find a meal. Loons begin calling as they arrive at their summer nesting areas. No other sound proclaims so eloquently that spring has arrived in the North Woods as the wail of a loon at dawn.

By late June, around the time of the summer solstice, blue flag irises are blooming in the beaver ponds and meadows, and the weather suddenly turns balmy. Summer flowers are not found in the forest, as the canopy cover is now complete and the understory is in deep shade. Instead, open beaver meadows are now the stages for spectacular blooms of asters, goldenrods, milkweed, joe-pye weed, and fireweed as well as Canada bluejoint grass and numerous sedges. Plant communities in drained beaver meadows are the North Woods' version of a prairie.

Summer in the North Woods is primarily the season of berries. Strawberries ripening in rocky openings begin the berry season in July. Their intense smell, especially when mixed with the resinous tang of pine needles atop sun-warmed bedrock knolls, can be almost overpowering. They are followed in quick succession by raspberries, blueberries and huckleberries, juneberries, and in September, especially in the Maritime Provinces, lingonberries.

Autumn creeps upon us before we and the animals are prepared for it. Snow flurries are not unheard of during the last week of August, although the first killing frost will not happen for a month. Bears are fattening on the berries and fruits, often walking the stems of juneberries and plums to the ground between their legs as they graze the ripe fruits to put on fat for the winter. Chipmunks and red squirrels are packing away the calories of seeds and nuts from conifer cones and hazels. Bees are completing the stores of honey they need to survive the coming cold. Teals have finished their fall migration by the end of August, and waves of mallards and other puddle ducks are beginning to pass through. Beavers are felling aspen along the shores of their ponds and dragging the branches, often still with their leaves, to their winter food caches beside their lodges, which will soon be encased with ice.

The greens of the deciduous foliage begin to fade as chlorophyll production ceases in early September, allowing the brilliant yellow, red, and orange carotene, anthocyanin, and xanthophyll pigments to emerge against the clear, deeply blue skies. The colorful death of the leaves is a programmed death known as abscission: A layer of cork forms at the base of the leaves' **petioles**, and nitrogen, phosphorus, and other nutrients are retracted from the leaf blades and stored in the twigs to supply the bursting of buds next spring. As the nutrients are moved from leaves to twigs, moose switch their foraging strategies from stripping twigs of their salad of green leaves to browsing the twigs themselves, which have suddenly become a tad more nutritious. By the third week of October, these leaves have all been shed. One day, the color is overhead and the next day it is underfoot, bringing the remaining nitrogen and phosphorus with it and completing the cycles of nutrients begun in spring by the bacteria, fungi, and actinomycetes.

In the North Woods, the fall equinox arrives after autumn's biological processes are finished. Winter is at the doorstep. The ground may remain bare of snow for an additional month after the leaves fall, or snow may follow soon after. The first snowfall will probably not be the first one of the winter-long snow cover, but continuous snow cover is usually not far behind. Cold polar air begins to arrive. Within a few weeks of the fall equinox, the forest has gone from a riot of color to a sober world of dark conifers against the white snow. Only the skies remain blue.

The Complexity of the North Woods

Eighteen thousand years ago, the area currently occupied by the North Woods was covered by an immense ice sheet centered on Hudson's Bay. As it retreated, the ice sheet left a diverse landscape of morainal ridges, flat outwash plains, and shallow glacial lakes. The moraines and outwash plains were open seedbeds to tree species expanding their ranges northward as the climate warmed. Many of the species that characterize the North Woods today were far south of the margin of the ice sheet eighteen thousand years ago. The North Woods came into being as the ranges of these species merged together in the new northern landscape.

Approximately fifty to sixty tree species in total make up the entire North Woods today. Their ranges stretch the entire east–west length of the biome with only a few exceptions. A hectare (10,000 square meters, or about 2.5 acres) of North Woods could encompass fewer than a dozen tree species, sometimes only three or four. The same three or four species may dominate the adjacent hectares. In contrast, a hectare of the Amazon rain forest might harbor several hundred tree species. Unlike the North Woods, the adjacent hectares in the rain forest may also contain several hundred tree species, but it is likely that most of them would be different species. This large number of tree species with small ranges is responsible for the complexity of the rain forest ecosystem.

The complexity of the North Woods lies not in the number of species, as in the Amazon rain forest, but in the very different ways each species gathers energy, food, and nutrients, interacts with other species in the food web, and alters nutrient cycles. There are fewer species in the North Woods than in tropical forests, but each species does something very different from the rest. Consequently, changes in the abundance of one species here very quickly translate into large changes in the ecosystem and landscape.[2]

The different shapes of leaves and crowns are important, and obvious, differences in the ways North Woods species capture light and turn it into sugars. The leaves of the thousands of tropical tree species are almost all ovate with smooth margins. In contrast, the diversity of leaves of North Woods trees includes single needles in spruce, hemlock, and balsam fir; multiple needles of pines and tamarack, which are connected into a bundle at their bases; single ovate leaves with serrated edges in beech, aspen, and birch; compound leaves with ovate leaflets with smooth edges in ashes; and lobed leaves with serrated edges in red maple or without serrations in sugar maple and red oak. Evergreen crowns range from the sharply conical balsam fir, the candle-like black spruce, the conical white and red spruce with clubby rather than pointed apices, the broad and spreading crowns of red and jack pine, and the elegant candelabra crown of white pine. Deciduous crowns are globular and may have their leaves dispersed throughout the crown, as in aspen and birch, or in a shell along the perimeter, as in maple and red oak.

Different leaf lifetimes have major effects on the dynamics of carbon uptake and nutrient cycles. Evergreen conifer needles can live for 2 years to a decade, whereas deciduous leaves (or tamarack needles) live for only a year. These different lifetimes determine how long atoms of carbon or nutrients reside in the needles or leaves. During their residence in the leaves, nutrients are temporarily delayed from cycling through the rest of the ecosystem.

Other important components of the complexity of the North Woods are the feedbacks between two or more species or between species and the nonliving components of the environment, such as water and nutrients. A feedback loop consists of two or more components that cycle energy, water, nutrients, or information in one direction around the loop. The components of a feedback loop therefore affect one another through these cycles. You are familiar with one common feedback loop, which is the thermostat, furnace, and air in your house. The thermostat turns on the furnace when the air temperature falls too low, the furnace heats the air, and the warmer air turns off the thermostat. Thermostat, furnace, and air exchange information through the electrical signal from the thermostat to the furnace, the input of heat from the furnace to the air, and the heated air shutting off the thermostat.

Feedbacks come in two flavors, negative and positive. Negative feedbacks happen when an increase in one component decreases the activity of another. Negative feedbacks dampen small changes in the two components of the feedback loop and thus stabilize the ecosystem. The thermostat–furnace–air feedback loop is a negative feedback that keeps the temperature of your house stable because the heating of the air by the furnace turns off the thermostat and shuts down the furnace. In nature, predator–prey feedbacks are usually negative feedbacks because growth of the prey population provides more food for the predator, but the growing predator population then kills and consumes more prey, thereby dampening the initial growth of the prey population. Predator–prey feedbacks are important regulators of the stability of almost every ecosystem.

Positive feedbacks between the two components of the feedback loop, on the other hand, reinforce or amplify a small change of one component, thereby often destabilizing a system and moving it toward another state. Engineers usually try to prevent positive feedbacks from happening in most appliances or machines because we want them to perform in a consistent, stable manner. But positive feedbacks are common in natural ecosystems and are important processes that control the recovery of ecosystems from disturbances such as fire or wind. For example, the resinous, flammable needles of jack pine promote fires; hotter and more frequent fires open the cones of pine, disperse their seeds, and kill the maple and birch competitors of the seedlings. The dispersed pine seeds then germinate in the burn, and the abundant seedlings grow into nearly pure pine stands free of competition by maple and birch. The resinous and flammable litter of the pine accumulates until the next fire, which begins the process again, keeping the stand in a nearly pure conifer state. The mutually promoting positive feedback between fire and jack pine moves the forest from mixed-species stands of deciduous maple and evergreen conifers toward pure stands of jack pine.

These positive and negative feedbacks between different species and fire raise interesting and unresolved questions about the role of fire in northern ecosystems. The classical idea of ecosystem recovery from disturbances such as fire that open the canopy is that the land is first colonized by pioneer species such as aspen and paper birch, which can grow rapidly in full sunlight. However, their seedlings cannot grow as well in their shade as the more shade tolerant maples, yellow birches, basswoods, or beeches. Eventually, the understory maples, yellow birches, basswoods, and beeches grow into the canopy and succeed the aspens and birches as the latter near the end of their lives. Each generation of maple, yellow birch, basswood, or beech is replaced by the next, which is lying in wait in the understory. This is the so-called climax forest, which persists until its canopy is opened by some exogenous disturbance. But how should we view the role of fire in plant succession in the North Woods? Clearly, fire is a disturbance when it kills climax maple forests. But what about pine forests? Is jack pine a pioneer or climax species? Is fire a disturbance to jack pine forests or an essential part of them? Maybe we have to think about fire in different ways for different parts of the North Woods.

As in the Arctic tundra,[3] population swings of very large amplitudes are other important manifestations of the complexity of the North Woods. When these swings in populations happen on a regular periodic basis, they are known as population cycles. For example, periodic outbreaks of spruce budworm defoliate spruce or fir every 30 to 60 years in northern conifer forests; similar outbreaks of forest tent caterpillar defoliate aspen every 10 to 15 years. Once defoliated, the spruce and fir die, but the aspen can recover and produce another crop of leaves, although its growth is slowed during years of tent caterpillar outbreaks. A population cycle in one species (the insects) can therefore drive population cycles in other species (their plant hosts), resulting in a feedback loop between them. Although much research has been done on the causes of these population cycles, their causes are still not fully understood.

Feedbacks between species during their mutual population cycles also govern the coevolution of both species. Tent caterpillars and aspen are a classic example of coevolution between an herbivore and its plant host. When tent caterpillars defoliate aspen, the aspen counters by producing noxious compounds to deter the caterpillars. Individual caterpillars that can tolerate these noxious compounds are favored in the proverbial "struggle for life" (a classic example of Darwinian natural selection). An evolutionary arms race develops out of this positive feedback: An outbreak of tent caterpillars drives the aspen population to evolve toward individual trees that can better defend themselves, driving the selection of the next generation of caterpillars toward those that can overcome these stronger chemical defenses, which results in a further selection for even more noxious compounds in the next generation of aspens.

A species is not at the same point in its cycle across the entire landscape at the same time. For example, spruce budworm can be abundant and killing spruce and fir at the peak of their cycle in one area but at a low point a few tens of kilometers away. After a spruce budworm outbreak, the mature stand is replaced by one composed of younger trees, but the stand without a spruce budworm outbreak continues to age unmolested. These asynchronous outbreaks of spruce budworm across the landscape create a complex mosaic of patches of spruce and fir of different ages. Each age class provides habitat for different bird and mammal species. Because of the wide population cycles of many northern species, the North Woods is not, and has never been, a uniform expanse of old growth climax forests but a complex mosaic of species that are in different stages of adjustment to the presence and abundance of the others.

The very different ways that North Woods species interact with each other and their environment make it easy to do experiments to test hypotheses on the role of a particular species in the food web: Simply remove or add a species and see what happens. To find out how moose affect the ecosystem, fence a large section of the forest off from moose and watch how the forest and soils change over the next several decades.

To find out how a particular tree species affects soils, harvest it from a forest (or alternatively establish pure plantations of that species) and monitor changes in soil chemistry. But these experiments often have to be done over large areas to encompass a representative portion of the entire ecosystem. They therefore often have the same character as forest management practices. It is not surprising that much of our knowledge of the ecology of the North Woods comes from large-scale and long-term experiments in National Forests in collaboration with forest managers and loggers.[4]

The North Woods played a major role in the development of the modern quantitative science of ecology. Questions about the natural history of the North Woods helped us develop our modern ideas of forest succession, the role of forest fire, responses of forests to climate change (both past and future), how watersheds work, how herbivores such as moose and beaver engineer their habitat, the relationship of a predator to its prey, how species avoid competition, the structure of food webs and the flow of energy through them, and many other ideas. We are just now beginning to learn how some of the natural history of species and the feedbacks we discussed earlier link them into food webs and ecosystems. But many questions remain open. These open questions are the grist for tomorrow's research by today's undergraduate and graduate students.

The North Woods and the American Idea of Nature

Besides being the source of many of our ideas on ecology, the North Woods is the theater on which much of the American character has emerged while we have been extracting resources from it. Humans, especially Europeans, have been and are now a major shaper of the North Woods.

On the western shore of Lake Superior at the border between Minnesota and Ontario is a set of rapids around which the voyageurs lifted their canoes and cargo along the Grand Portage on their way to the forests of the North Country in search of beavers. The birth of natural history in North America can be found in the journals of these fur-trading explorers, such as Samuel Hearne, Alexander Mackenzie, and most especially David Thompson. Thompson's descriptions of the forests between Lake Superior and Hudson's Bay west to the prairies, often written from his canoe as he traveled his beaver trade routes, contain ecological questions that even today are unanswered. This immense land was known to the voyageurs as *le pays d'en haut* (The Upper Country) or *le pays du Nord* (The North Country), and it was in *le pays* where the fur trade held sway for close to two centuries. The fur trade inserted Europeans squarely into the ecology of the region and almost caused the extinction of beaver, moose, lynx, wolves, and other species.

The French and Indian Wars were fought throughout the North Woods from Maine to Lake Superior, partly in order to determine whether the British or the French controlled the fur trade. This war brought soldiers farther into the North Woods. Several serendipitous discoveries in natural history resulted, most notably the discovery, in 1705, of the first mastodon tooth in upstate New York by French soldiers on patrol. Unlike fur traders and voyageurs in canoes, an army needs trails and roads to transport ordnance and supplies. Roads surveyed during the French and Indian Wars later brought waves of settlers into the North Woods, where they cleared the forest and brought some of the first exotic invading species from Europe, such as St. John's wort.

Our presence in the North Woods continued with the cutting of the large white and red pines, beginning in New England. By the mid-1800s most of the easily harvestable pine in New England had been taken. The loggers then moved farther and farther west, removing the pine as they went. The pine harvest, and that of other species once the stock of pine dwindled, set into motion a chain of natural events that continues today, including changes in the fire regime.

Logging left much slash covering the landscape. This slash was then fuel for large and very intense fires. Unlike most natural fires, these fires in logging slash were so hot that they burned the topsoil away. Bereft of its protective cover of forest floor, the soil eroded down to bedrock in many places. This exposed veins and beds of copper and iron ores, and mining soon followed. Extraction of iron and copper forever altered entire landscapes in the Adirondacks, Ontario, the Upper Peninsula of Michigan, and northern Minnesota.

The harvesting of timber and furs and extraction of iron and copper from the North Woods were the economic foundations for the growth of cities near its southern edge, such as Duluth and Detroit, Chicago and Cleveland, Milwaukee and Minneapolis, and Montreal and Toronto, among many others. At the same time as these growing cities were being supplied by resources extracted from the North Woods, a few naturalists began sounding the alarm for its future. Their writings were the origin and development of the ideas of wilderness preservation and land conservation. Peter Kalm, a colleague of Linnaeus who traveled throughout the North Woods and collected specimens, noticed the demise of the North Woods over large areas of New York even before the Revolutionary War.[5] A half century later, as he paddled up the Allagash River in Maine to its source at Heron Lake, Henry David Thoreau wrote several essays that were later compiled into *The Maine Woods*, one of the most literary descriptions of the natural history of the North Woods.[6] The contrast between what Thoreau saw on this trip and the lack of any sizeable original remnant of the North Woods in Massachusetts may have been one of the inspirations for him to pen his famous saying, "In wildness is the preservation of the world." Thoreau's *The Maine Woods* led George Perkins Marsh, a native Vermonter, to write *Man and Nature*, in which he proposed that intact forests stabilize streamflow, perhaps the first attempt to provide a scientific rationale for land and water conservation.[7] Marsh's book led directly to the preservation of the Adirondack Preserve in 1892, reserving as "forever wild" all lands within the preserve's borders owned by the State of New York in order to protect the headwaters of the Hudson River.[8] The Adirondack Preserve lies entirely within the North Woods biome and, at 25,000 square kilometers, is the largest wilderness park in the contiguous United States, although not all of it is virgin forest. Other preserves in the North Woods followed over the next century, most notably the 4,400-square-kilometer Boundary Waters Canoe Area Wilderness in northern Minnesota, of which more than half is uncut virgin forest, the largest such contiguous tract in eastern North America. The existence and borders of the Boundary Waters would not be possible without the efforts of Bud Heinselman, an ecologist with the U.S. Forest Service who, harking back to David Thompson, mapped these forests from his canoe and in the process discovered the history and role of fire in maintaining the diversity of northern forests.[9] Clearly, the development of the idea of wilderness and land conservation owes much to the naturalists and scientists who unraveled the natural history of the North Woods.

The wilderness areas of the Adirondacks, the Boundary Waters and adjacent Voyageurs National Park and Quetico Provincial Park, Isle Royale National Park, and other preserves are, in Aldo Leopold's words, land laboratories[10] where we can further investigate the natural history of the North Woods and how it works as an intact ecosystem. These wilderness areas have preserved large tracts of the North Woods remarkably well. But climate change, which fostered the assembly of the North Woods after the retreat of the ice sheet, may in the next several decades foster its disassembly. Just as the North Woods did not migrate intact from the south after the ice sheet melted, it will not simply migrate intact in response to a changing climate in the near future. Rather, individual species will respond to a changing climate in ways determined by their natural history. To preserve the North Woods for the future, it will not be enough to simply set aside large chunks of preserves, valuable as these continue to be. We will need to preserve the climate of the entire planet. If we do not, new assemblages of species will form and the North Woods as an intact ecosystem will probably almost, if not entirely, disappear. We are on the cusp of whether we want to try to stop the worst of this.

We are now responsible for the future of the North Woods. Sound local, regional, and global environmental policy will need to be made on a foundation of solid understanding of the natural history of the North Woods and all biomes and ecosystems everywhere. Much of this natural history is known, but many questions remain unanswered for the North Woods and every other biome. In the chapters that follow, we will see that the North Woods is a fount of natural history questions that can point us in new directions in ecology, evolution, geology, and conservation biology. So let us begin.

PART I

The Assembly of a Northern Ecosystem and the European Discovery of Its Natural History

If you fly from Minneapolis or Chicago to northern Europe, get a window seat on the north (left) side of the plane (ecologists and geologists should always ask for window seats). Watch the landscape of northern Minnesota, Wisconsin, Ontario, Maine, and Labrador pass by instead of watching the movie (trust me, the landscape is better). In many places and for many hours, you will see more water than land. Rivers, sometimes braided, meander across the landscape. The lakes will be in every conceivable size, from small ponds to large inland seas, and in every shape, from almost perfectly round holes that look like kitchen kettles to wormlike lakes with long axes aligned north–south. The lakes are the eyes of the landscape. The sun flickers off their surfaces. Picture yourself down there, traveling through it. In summer, it is easier to canoe through most of the northern landscape than to walk across it; in winter, the frozen and snow-blanketed lakes and rivers are level and open highways for travel by snowshoes, skis, and dogsleds.

The abundance of water in all its forms—snow, ice, liquid water, and vapor—defines the northern landscape just as the absence of water defines the desert. Why does this landscape hold so much water? What determines its distribution and flow? How does the water affect the life cycles of the plant and animal species that make this northern ecosystem their home? How do the plants and animals in turn determine the distribution of water?

The assembly of this northern ecosystem is the story of how plant and animal species moved into a landscape sculpted by a massive ice sheet and how these species responded to and modified the distribution and flow of water in all its forms.

1.
Setting the Stage

How the ice sheet sculpted a landscape of surprisingly high diversity of landforms that control the flow and distribution of water.

Robert Frost, whose poems often contain perceptive observations about the natural history of the North Woods, said that ice "would suffice" for the world's destruction,[1] but he had it exactly backward. Ice, in the form of a massive continental ice sheet, was responsible not for the destruction of the northern world but for the creation of the landscape on which the North Woods assembled itself. Ecology may be the theater for the evolutionary play,[2] but the sculpting of this landscape by the ice sheet set the stage for the assembly of northern food webs and ecosystems and the evolution of the organisms they comprise. Before the North Woods assembled itself as its species arrived from the South, before the moose, beaver, and loons arrived, there was the land emerging from beneath several kilometers of ice. The natural history of organisms is in part a set of adaptations to the landscapes they live in, and these landscapes are shaped by their underlying geology.

Our understanding of how ice created the northern landscape did not begin in North America but along the Swedish shore of the Gulf of Bothnia in the Baltic Sea, a region whose forests and shorelines are strikingly similar to those of Lake Superior, both today and in the distant past. In the early 1700s Anders Celsius, a Swedish geographer, astronomer, colleague of Linnaeus, and the inventor of the temperature scale, was studying records of the locations of old Viking villages. Celsius noticed that quite a few villages that were on the shore in Viking times were by then up to a kilometer inland. Being seagoing people, Vikings did not move villages inland. Plenty of people, from local villagers to scholars in Uppsala, must have noticed the inland displacement of Viking villages but gave it no further consideration. Instead, it was Celsius who thought long and hard about what this observation might mean. His thoughts and investigations eventually led to a revolution in the way we understand the earth in general and the origin of the North Woods in particular.

Celsius realized that the inland displacement of Viking villages meant that either the land was rising or the sea was falling, but he did not know which or why. To shed more light on what was happening, he decided to measure the rate of this displacement by taking a chisel in hand and carving lines in cliff faces at the water surface on the Baltic shore in northern Sweden. Above the line, he carved the date and his initials. He then instructed his students and their academic descendants to do the same periodically and measure the distance between his line carved in the early 1700s and theirs carved decades and centuries later. Thus began what is perhaps the first systematic, long-term ecological experiment.

These chiseled marks, which can still be seen today, are aligned in columns up several rock faces on the northern shores of the Baltic. They remind one of the marks parents sometimes make on doorjambs to record the growth of their children. They are often accompanied by a crown with the initials and number of whichever King Gustav happened to reign at the time. These are striking hieroglyphics, with deep scientific meaning about the history and stability of the land. I've visited these shores several times during the past two decades while doing research in northern Sweden with my friend and colleague Kjell Danell. It is always a remarkable experience to stand at the waterline, look at these marks, and realize that the spot where Celsius knelt and made them three centuries ago is now 2 or 3 meters up the rock face. These marks were the first demonstration that sea level and the crust of the earth are not stable; thus began the science of geology as the study of the dynamics of the earth and not just its structure.

Celsius eventually came to think that the vertical displacement of land and sea meant the sea level was falling. Given a long enough record, this hypothesis could be easily tested. If the land is stable and the sea level is falling, then the sea level should fall down from Celsius's original marks by the same amount at every spot along the coast. But by the late 1700s, his students found that the vertical displacement of land and sea was uneven across the Baltic. Swedish geologists then concluded that the land must be rising and taking Celsius's original chisel marks with it at different rates in different places. Still, they did not know (yet) what was causing the land to rise.

The next clue to what was happening came from Switzerland. In the early 1800s, Jean Charpentier, a Swiss geologist, called the attention of Louis Agassiz, his compatriot and colleague, to the boulders scattered across the Swiss landscape downvalley from glaciers. These boulders were curious because they were of different rock types from the local bedrock they were perched on. Similar boulders had been noted throughout northern Europe from Scotland to Finland and east into Russia. Clearly they had been transported there somehow, but the sizes of many of these boulders, from that of a cabbage to a cottage, raised the obvious question as to what the transporting agent could be. It had been thought that they were transported to their present locations by icebergs drifting on Noah's worldwide flood, so these rocks were called drift, or erratics, from the Latin *errare*, meaning "to wander or to roam far from home." But Charpentier and Agassiz realized that the Swiss boulders had clearly been transported by the mountain glacier because they could see them dropping from the face of the glacier like bowling balls.

In a bold creative leap, Agassiz extrapolated from the glaciated Swiss valleys to all of northern Europe and proposed, in 1840, that a giant **ice sheet**, or continental-scale glacier, once covered the northern half of the continent and transported these boulders southward.[3] Later, after he moved to Harvard and explored the landscapes of Maine and Lake Superior, Agassiz proposed that a similar ice sheet once covered the northern half of North America.[4] Most geologists and naturalists at the time thought this was lunacy (except for Henry David Thoreau, who, in his *Journals*, provided a remarkably accurate interpretation of the landscape around Walden Pond in light of Agassiz's theories[5]). But Agassiz prevailed by the weight of the observations and his eloquence (Agassiz's writings are some of the most lucid of all natural history writings). By his death in 1873, most reputable geologists and naturalists had accepted his theory of the Ice Age in principle, although the details still needed some working out.

In 1865, shortly after Agassiz's original hypothesis of a continental ice sheet, Scottish geologist Thomas Jamieson proposed that an ice sheet thick enough to cover a continent must have been heavy enough to warp the crust of the earth downward into the more viscous mantle beneath it, pushing the mantle outward. (The mantle is the layer of hot but not molten rock between the lighter crust floating atop it and the molten core beneath it. The consistency of the mantle is similar to that of a candle softened in the hot sun.) The Swedes had the last word, as Jamieson's hypothesis was confirmed for the Swedish coast in 1890 by Gerard De Geer from the University of Stockholm.

Now imagine a sheet of ice 2 or 3 kilometers thick (about 1.2 to 1.8 miles) covering the land from Minnesota to the Atlantic coast and north through Hudson's Bay 18,000 years ago. The ice age that produced the Laurentide Ice Sheet is known as the Wisconsinan Glaciation, the latest of at least four such glaciations in the past million years or so, each separated from the others by warm periods.[6] (The Wisconsinan Glaciation is so named because of the especially well-preserved glacial landforms in that state from this particular advance of the ice sheet.) These glaciations and the intervening warm periods were partly caused by wobbles in the earth's rotation and orbit; these wobbles determine the distribution of sunlight across the earth's surface and through the seasons. When the tilt of the axis and the shape of the orbit were aligned to minimize the amount of energy being delivered to the northern hemisphere in summer, then summers were cool enough that snow laid down during the previous winter did not melt, and the snow cover began to accumulate.[7] Year after year, decade after decade, the snow accumulated and compressed the lower layers into ice. When the ice accumulated to 50 meters thick, its weight caused it to stop being the brittle substance we are familiar with. The ice sheet was born and began to ooze southward across the continent.[8]

Ice continued to accumulate in the thick core of the ice sheet even as it oozed outward along its thinner margin. The weight of the ice in the thicker core ensured the continued flow of ice downslope to the margin. As the ice sheet flowed south it entered a warmer climate where melting at the ice sheet's surface exceeded snowfall. The front of the ice sheet had passed out of its **zone of accumulation** at the core and entered its **zone of ablation** near the margins. The ablation zone is where there is a net loss of ice by melting, snow blowing off the surface, and calving from the **snout**, or front, of the glacier. The ice in the ablation zone can be sustained only by continued flow from the core. Where the rate of melting and calving at the snout of the ice sheet equaled the inputs from both upglacier flow and local snowfall, its advance came to an end. This point defines the **terminus** of the glacier. (The snout is the physical "nose" of the ice sheet, and the terminus is its location. You stand at the terminus but touch the snout.)

At its greatest extent, the weight of several kilometers of ice—nearly 2 metric tons above every square centimeter—warped the earth's crust downward 80 meters near the ice margin and more than 400 meters beneath its thick core in Labrador, east of Hudson's Bay. Even today, the crust beneath the Greenland Ice Sheet is still warped nearly 400 meters below sea level.

About 15,000 years ago, continued changes in the wobbles of the earth's rotation and orbit moved the earth to a position where the energy from the sun reaching the northern hemisphere was enough to melt the winter's snow during the next summer. When snow accumulation in the core decreased, flow of ice to the ablation zone slowed. At the same time, warming also accelerated melting in the ablation zone. Slower flow of ice downglacier and faster melting in the ablation zone caused the terminus to retreat northward. The ice sheet then began to disintegrate at its southern margin, opening up land for the northward expansion of the ranges of plant species. Except for remnants such as the Greenland Ice Sheet and a few small glaciers remaining on Baffin Island and other high Arctic islands, the last vestiges of the Laurentide Ice Sheet had melted by 9,000 or so years ago.

As it was relieved of its burden of ice, the crust rebounded rapidly, much like a trampoline, then more slowly as the mantle oozed back underneath. For the past 10,000 years since the ice sheet retreated from the current region of the North Woods, the crust has continued to rise near Lake Superior, Hudson's Bay, and in northern New England, especially along the Maine coast.[9] Along the shore of Hudson's Bay and along the Gulf of Bothnia in Scandinavia, the land is still rising slightly more than 1 centimeter per year, as Celsius's chisel marks first showed. The rebound of the crust and redistribution of the mantle have been sufficiently large to change the shape of the earth and alter its rotation.[10] The rebound is not yet over and won't be for another 10,000 years or more.

After it melted, the ice sheet left behind a watery terrain as its legacy. The unequal rise of the crust during the rebound controls much of the regional distribution of water in the northern landscape. The land in the north is rising faster than that in the south, partly because of the greater thickness of ice near the ice sheet's core than at its margin and partly because the northern lands were relieved of their ice burdens much later. The greater rise of the earth's crust in the north tilts the land's surface toward the south and spills water onto the southern shores of the Great Lakes in North America and large lakes in Scandinavia.[11] The mouths of rivers flowing to the southern shores of the Great Lakes are drowned by water spilling into them from the faster-rising northern shores. Rivers along northern shores, in contrast, flow straight and fast into lakes, often down precipitous waterfalls marking earlier shorelines.

If the rebound of the land from the weight of the ice determines the regional distribution of water in this northern landscape, then the sculpting of the landscape by the ice sheet—carving the bedrock here, depositing debris there—created the topographic and geologic template on which today's watersheds are organized.

This topographic and geological template was created by the down-glacier flow of ice, which was a giant conveyor belt that continuously plucked clays, silts, sands, cobbles, and boulders from the previous landscape and either brought this unsorted debris forward to the snout or plastered it on the slopes of hills and in the valleys. This unsorted debris of dirt, cobbles, and boulders is collectively known as **till**. The cobbles and boulders in the till were often transported on a journey many hundreds of kilometers from their origin and, when deposited atop different bedrock from that where they originated, became Agassiz's erratics.

The ice sheet sculpted the landscape in three ways: by carving the bedrock, soil, sediments, and debris beneath it; by bringing the till forward and depositing it at its snout; and by releasing meltwater that washed, sorted, and deposited materials across the landscape in front of it. These landforms are the record of the climate history of the glaciated regions of the earth for the past 18,000 years.

Beneath the flowing base of the ice sheet, the sand embedded in it smoothed and polished the underlying bedrock. Rocks embedded in the ice carved parallel grooves onto the surface of the underlying bedrock in the direction of the ice flow. Whenever I stand on top of one of these polished bedrock surfaces and look down the axes of these grooves, I almost feel the grinding power of the ice sheet at my back. In places, the rocks in the ice chattered as they were dragged forward, making nicks on the bedrock surface where flakes were chipped off. These nicks are known, appropriately, as chatter marks. Smaller grooves and chatter marks were also carved on most of the cobbles and boulders embedded in the till.

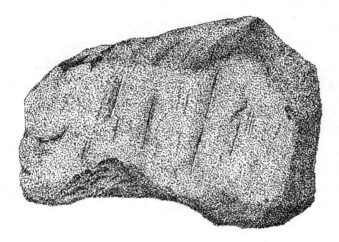

The conveyor belt of ice brought the till forward and deposited it in a broad ridge known as the terminal moraine[12] at the southernmost terminus of the ice sheet. Cape Cod, Nantucket, and Long Island form a long sweep of terminal moraine from the Wisconsinan Glaciation out into the Atlantic. The terminal moraine then extends westward through New Jersey, Pennsylvania, Ohio, Indiana, and Illinois before veering northwest into Wisconsin. It then arcs northwest through Minnesota, Manitoba, and beyond to the Canadian Arctic, running along the edge of the great granitic Canadian Shield just west of Lake Athabasca, Great Slave Lake, Great Bear Lake, and thousands of smaller lakes in between, like beads on a rosary. Except for a finger pointing down the spine of the Appalachians, the North Woods lies entirely inside the continental arc of the terminal moraine.

I've been speaking of the terminal moraine as if it were a single long ridge deposited all at once, but it's a bit more complicated than that.

The ice sheet was actually composed of a number of lobes flowing more or less southward, each depositing its own debris, including its own terminal moraine. Each moraine has its own signature till determined by the bedrock eroded by the parent lobe that gave birth to it. The lobe that deposited the moraines closest to my home in northern Minnesota is called the Superior Lobe because much of the material was scooped out of the bottom of Lake Superior by the ice sheet. What I have been calling the terminal moraine is actually a system of terminal moraines made by each of these lobes.

As the climate warmed, melting at the snout exceeded the forward flow of ice, and the terminus began retreating. When the ice was advancing, the snout was a tall, straight cliff of clean ice, like the bow of a ship plowing through the sea. But as the ice sheet melted, crevasses were cut into the ice down which waterfalls often flowed. The once tall, straight, and clean snout became rounded and furrowed by the melt-water. Although the terminus of the ice sheet retreated northward, the flowing ice would continue to bring till south to the melting snout, which became dirty in its lower layers.

The till was dumped in new moraines that form broad arcs roughly concentric to and inside the terminal moraine. These moraines are called end moraines, and some of them are large enough to form major divides of water flow across the landscape. The massive Nickerson moraine, formed by the Superior Lobe about 12,000 years ago 70 kilometers south of Duluth, is 6 kilometers wide and divides water flowing north to Lake Superior and eventually to the North Atlantic from water flowing south to the Mississippi River and eventually to the Gulf of Mexico.[13] In Minnesota, the North Woods lies almost entirely north of the Nickerson moraine.

There are other types of moraines that are somewhat flatter but still rolling expanses of till. These are of two types. Ground moraines were deposited beneath the ice sheet as it advanced and were then exposed during a rapid retreat of the terminus, when the snout did not stay long enough in any one spot to build a new end moraine. Stagnation moraines were deposited when the ice behind the snout had melted away to the point where it was no longer thick enough to flow. The ice simply stagnated in place, and the till in it was dropped as the ice broke into pieces.

Isolated blocks of ice in stagnation and end moraines were often covered by the till as it was melted out from the uppermost layers of the ice sheet. Insulated from the rising air temperatures by their blankets of till, these ice blocks melted slowly, taking a millennium or more to completely disappear. As the buried ice blocks melted, the surface of the till above them sagged and collected overland flow from rains and melting snow. This water percolated downward to the buried ice block and accelerated its melting. The continued melting of the ice blocks and slumping of the overlying till into the sag formed round bowls with steep sides reminiscent of large kettles. As the bowls deepened, they eventually intersected the regional groundwater surface and filled permanently with water, forming what are now known as **kettle lakes**. The most famous kettle lake, and the deepest in New England, is Thoreau's Walden Pond,[14] which has two basins, each probably formed by two adjacent blocks of buried ice. Swarms of kettle lakes sweeping in broad arcs across the landscape can be seen on many road maps of glaciated areas.

Other lakes, known as glacial lakes, were filled by meltwater trapped between the retreating snout and an end moraine some distance in front of it. Many of these glacial lakes eventually disappeared as drainage channels gradually opened up. Deeper glacial lake basins either contain a small remnant of the original lake, such as Red Lake in Minnesota, or else contain the descendant of the river that drained them. Decaying vegetation along the shore and from floating bog mats sometimes filled these basins with partly decayed organic matter known as peat. The wetlands that now fill these basins are known as peatlands and are occupied by a unique set of communities composed of sphagnum mosses, shrubs such as Labrador tea, leatherleaf, and bog rosemary, black spruce and tamaracks, and a variety of sedges, orchids, and other plants. The peat in these basins can be up to 3 meters thick. Half the weight of peat is carbon originally taken from the atmosphere by the peatland vegetation. Although northern peatlands cover only 3 percent of the earth's land surface, they contain one third of all the carbon in the world's soils.[15]

The ice sheet also shaped the till beneath it into elegant landforms that emerged as the ice sheet retreated. The two most common of these landforms are called eskers and drumlins. **Eskers** are the remains of the beds of streams that meandered through a tunnel underneath the ice sheet. When the ice sheet melted away, the graceful, sinusoidal shapes of the eskers were exposed. Skiing or snowshoeing along their flat tops is a joy compared with picking your way through the hummocky ground moraine that often surrounds them. Eskers can also be dry highways through the soggy peatlands now filling glacial lake basins.

Drumlins are teardrop-shaped hills, looking like huge overturned, half-buried spoons, a kilometer or two in length, 50 meters high, and a few hundred meters wide. Their long, gently sloping tails point in the downstream direction of ice flow. Drumlins rarely occur singly but are often clustered in herds of hundreds, known as a drumlin fields. They rise out of the sea of surrounding till like a pod of whales breaking the surface of the water. Drumlins in a drumlin field shed water off their shoulders, supplying reticulated networks of wetlands that look like a gill net draped throughout the field, the individual drumlins caught in the net like herring.

Sheets of meltwater washed clays, silts, and sands out of the ice sheet and across the landscape many kilometers in front of it. The sands were deposited in large aprons in front of the terminal moraine or draped over ground or stagnation moraines behind it, while the clays and silts were washed away. These broad, flat, sandy outwash plains are a striking contrast to the clay and boulder ridges and hummocky landscapes of moraines. Where enough sand was washed out of the ice sheet to bury isolated blocks of ice, kettle lakes also formed; where they did, the plain is known as pitted outwash. The most famous kettle lake in pitted outwash is Cedar Bog Lake in Minnesota, where limnologist Raymond Lindeman invented the concept of energy flow through **trophic levels** in a food web.[16]

The highly diverse set of landforms of moraines, outwash plains, drumlins, eskers, and glacial lake basins left by the ice sheet determines the distribution, abundance, and flow patterns of water in the soils, lakes, wetlands, and streams of the northern landscape. This watery landscape became the stage on which the North Woods assembled itself. Onto this stage marched the players, beginning with the plants. Mammalian herbivores such as beaver and moose as well as insects soon followed, forming new food webs that never before existed. These herbivores were soon followed by their predators, such as bears and wolves, and the North Woods ecosystem came into being. Nine thousand years after the ice sheet retreated, the voyageurs in their birchbark canoes followed the rivers and chains of lakes through this watery landscape, searching for beaver pelts and adventure, thus initiating the discovery and exploration of the natural history of this biome.

2.

The Emergence of the North Woods

The North Woods did not exist intact south of the ice sheet waiting to move north but was assembled bit by bit as the ranges of plant species shifted northward and distributed themselves along the gradients of soil moisture in the landscape left by the ice sheet.

In May or June, trees dust the landscape with a thin film of pollen. Smear a thin layer of Vaseline on some glass slides and leave them overnight under pine, spruce, birch, or maple trees whose flowers are shedding pollen. The next day, with a binocular microscope under 400 power (10 × eyepiece and 40 × objective), you can see the intricately sculpted shapes of pollen characteristic of each genus of trees.[1] Pollen grains of pine, spruce, and fir look like squashed footballs with two conspicuous air-filled bladders attached to them; the bladders often have a network pattern in their surfaces that aids in identifying them to the species level. These bladders help keep the grains aloft in the breeze, enabling them to disperse widely. In contrast, pollen grains of deciduous hardwood species are more rounded grains with surface patterns. Birch pollen is somewhat flattened, with three evenly spaced conspicuous pores that give the grain a triangular appearance. Maples are slightly fattened spheres, with three shallow furrows evenly spaced around the equator that meet at the "poles" of the grain. With care and with more sophisticated micro-scopes, it is possible to distinguish the pollen from different species in the same genus.

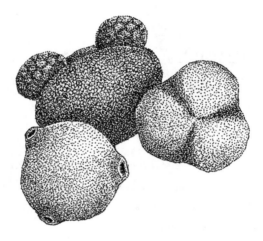

From the standpoint of the plant, most pollen grains that drift across the landscape are wasted because they do not fall on female flowers and fertilize them. Some of this wasted pollen instead falls on the surface of lakes or ponds, where it is easily held in the surface film of water. If you live near a quiet lake or pond, crouch low at its shore when the pollen is blowing and look across the surface. You will see pollen of different colors, mostly from the nearby forest but some from far away as well, swirled across the surface. When breezes break the surface ten-sion, the pollen sinks downward through the water column to become entrapped, year after year, decade after decade, in layers of sediment. Where the sediment has remained reasonably undisturbed for thou-sands of years in kettle lakes and glacial lakes or in the peatlands that filled their basins, these layers became pages in a book recording the assembly of the North Woods.

This book of pollen in the sediment contains the answers to questions such as "How did the North Woods we know today come about as the climate warmed and the ice sheet retreated? Was it an intact biome lying patiently to the south of the ice sheet, waiting for the land to open up? Or did the different species arrive bit by bit? If so, what was the sequence of arrival, and how long did it take for each species to arrive?" Understanding the answers to these questions may also help us predict how the North Woods may respond to climate change in the future so that we can prepare for it.

The story of the assembly of the North Woods biome was deciphered by palynologists through painstaking analyses of the first appearances of pollen from different species in thousands of cores taken from the kettle and glacial lakes north of the terminal moraine. Carbon-14 dating of the layer in the core where its pollen first appeared gives us a good approximation of the time of first appearance of a species in the region. After several decades of analyzing pollen cores from numerous lakes, palynologists realized, somewhat to their surprise, that the North Woods was not an intact biome just south of the ice sheet waiting for the land to open up. At the height of the ice age, the North Woods hardly existed anywhere. Instead, it assembled itself bit by bit as each species arrived. As new species arrived, the other species that had already established themselves adjusted to the presence of the newcomer.[2] Today's assemblage of species in the North Woods is a young biome, only about 6,000 years old, which is to say only slightly younger than the beginnings of village life and the first steps of humans toward civilization. For long-lived species such white pine and hemlock, 6,000 years is only fifteen of their lifespans.

Black and white spruce were often the first tree species to arrive on the land emerging from beneath the ice sheet because they were already present at the ice sheet's margin. Pollen records from many lakes document that spruce migrated north in two prongs grasping the ice sheet, one prong reaching north into New England and the Maritime Prov-

inces in the east and the other prong reaching north into Minnesota and
Manitoba in the west. But these two prongs did not expand northward
at the same rate. The eastern prong spread northward at 200 to 300
meters per year, and the western prong sprinted northward at a phe-
nomenal 2,000 meters per year.[3]

How can spruce have dispersal rates that differ by an order of magnitude on opposite ends of its range? Squirrels can disperse seeds as they
carry them to winter caches, but squirrels have small home ranges and
could not have carried these seeds so far in short periods of time as in the
western prong. Crossbills eat spruce seeds as they open the cones and so
hardly disperse the seeds any distance at all. However, spruce seeds also
have winged keels that can catch the winds. Using a computer simulation model, J. E. Kutzbach and P. J. Guetter found that the steep temperature gradient between the ice sheet and the land to the south created
a strong stationary high pressure system centered over the ice sheet.[4]
This high-pressure system generated strong winds circulating clockwise.
Along the eastern edge of the ice sheet, these winds blew south, obstruct-
ing the northward migration of spruce in New England and the Maritime Provinces. Along the western edge, the clockwise circulation turned
these winds northward. These winds in the west carried the winged seeds
of spruce aloft, depositing them farther north, where they became the
first generation of spruce invading the newly opened land. After 30 years
or so, when these pioneers matured and began to produce abundant
crops of cones, some of their seeds were lofted farther to the north. By
overlaying the dates of first appearance of spruce in both prongs on
Kutzbach and Guetter's maps of postglacial wind circulation, Canadian
palynologists J. C. Ritchie and G. M. MacDonald showed that, around
9,500 years before the present, spruce migrated north 2,000 kilometers
from southeastern Alberta to the Mackenzie Delta within a short period
of 1,000 years. After that sprint of spruce northward, the rapid melting
of the ice sheet weakened the circulation of the high-pressure system.

Further influx of spruce in the west then proceeded at a more stately pace comparable to that on the eastern prong.

During the ice sheet's farthest advance 18,000 years ago, North Woods species other than spruce were in isolated refuges far to the south. White, red, and jack pine and hemlock were confined to refuges on the eastern seaboard in North Carolina. Balsam fir was found slightly farther north, in Virginia and Maryland, tamarack was found farther west of the Appalachians, in Tennessee, and oak, maple, beech, and chestnut were confined farther south, along the Gulf Coast.[5]

These species migrated out of their refuges into the deglaciated regions at different speeds and in a crisscross pattern of northeastward, northward, and northwestward movements, sometimes passing each other in different directions, sometimes leapfrogging over each other, and sometimes lagging behind one another. The pines migrated northwestward at about 350 meters per year, reaching northeastern Minnesota 7,000 years ago, and hemlock marched behind at a more stately pace of 200 meters per year, reaching western Wisconsin 2,000 years ago. Beech migrated at about 200 meters per year, at first northeastward up the Atlantic seaboard in a direction perpendicular to pine and hemlock. Then, about 7,000 years ago, for unknown reasons beech took a sharp turn west through the Great Lakes region. Although beech arrived in southern Michigan before hemlock, once hemlock arrived there it catapulted past beech, arriving in the Upper Peninsula 6,000 years ago while beech trailed behind, arriving 2,000 years later. Tamarack began migrating northeastward until about 1,000 years ago, whereupon it spread out along the Great Lakes and migrated northward along its entire front. Chestnut had the slowest migration rate, at 100 meters per year, with a straight and consistent northeast movement into New England. Only balsam fir expanded consistently northward across the Great Lakes region after moving out of Virginia, reaching the north shore of Lake Superior by 8,000 years ago.

The inconsistency in these migration patterns was a surprise to paly-nologists. When they began to describe pollen cores, palynologists did not have explicit hypotheses in mind about species migrations, but that doesn't mean that their conclusions about different rates and patterns of migration of different species are incorrect. These conclusions about different rates and patterns of migration simply emerged from the descriptions of a very large collection of pollen cores without any preconceived hypothesis in mind. Hypothesis testing is a very strong way to do science, but it isn't the only way. The perpendicular migration directions of pine and beech and the leapfrogging of hemlock over beech in Michigan could never have been predicted by any a priori hypothesis. Any hypothesis at this stage may have biased palynologists to consider only certain modes and patterns of dispersal but not others. The discovery of how the North Woods assembled itself and the understanding of how the northern landscape was created as the ice sheet retreated, as we saw in the previous essay, are excellent examples of doing science without explicit hypotheses during the initial stages of research.

But eventually, as we begin to understand the broad patterns of phenomena, we begin to form hypotheses about the factors that control these patterns. For example, the tree species of today's North Woods segregate across the landscape according to their drought tolerances and the wide-ranging water-holding capacities of the soils of different glacial landforms. You can see this as you drive across the landscape; in fact, the tree species present are often clues to the underlying glacial deposits.[6] Sandy outwash plains that hold little water within the rooting zone are home to the drought-tolerant jack pine and red pine. White pine can survive on outwash plains but also extends its ecological range onto the more silt- and clay-rich moraines that can hold more moisture. Moraines also support the drought-intolerant hardwoods such as sugar and red maples, basswood, northern red oak, and yellow birch, among other species. In the southern band of the North Woods, white spruce and balsam fir are found on the drier uplands, and black spruce is often confined to the peatlands occupying former glacial lakes. Toward the north, shorter and cooler growing seasons and consequently lower demands of the trees for water allow black spruce to creep out of the peatlands and across the upland.

Perhaps the soils on the glacial landforms controlled the migration patterns of species in the past as they control their distribution today. To test this hypothesis, Linda Brubaker studied pollen cores from three lakes, each on different types of outwash or till in the Upper Peninsula of Michigan.[7] If soil moisture controlled the assembly of different communities, then the pollen record should differ across these landscapes. If so, then the patterns of moraines and outwash plains emerging from beneath the ice sheet determined the assembly of the North Woods across the landscape.

Brubaker found that, for about the first 1,000 years after the ice sheet retreated, the landscapes around all three lakes was dominated by roughly the same forest of jack pine with small amounts of spruce and some birch. This jack pine–dominated assemblage persisted until about 9,000 years ago. At that time, the tilt of the earth's axis and the distance to the sun during the northern summer maximized the amount of sunlight impinging on the northern hemisphere.[8] Summers were warmer than today by 2°C and drier during this period, known as the Hypsithermal. These warm and dry conditions are best tolerated by jack pine. Brubaker found that during the Hypsithermal, spruce migrated westward out of Michigan into the cooler areas of Minnesota and then around Lake Superior into Ontario. At the same time, other warm-tolerant tree species migrated into the landscape but colonized different glacial deposits, depending on the amount of moisture in the soil and their tolerances of drought.[9] Drought-tolerant jack pine persisted on the outwash plains with the coarsest sands that hold little moisture, as they do today. White pine, somewhat less drought tolerant, became dominant on outwash plains with finer sand that can hold slightly more moisture. Sugar maple and northern red oak, even less drought tolerant, displaced jack pine completely on the clay-rich tills, although white pine became established there as well. This sorting of species along soil moisture gradients in the past has been confirmed by computer models that simulate climate, soil moisture availability, and responses of these species to drought.[10]

At about 5,000 years ago, the climate began cooling again. During this time, Brubaker found that spruce migrated back onto this landscape in northern Michigan. But as the climate cooled the remaining species did not leave in the same order or at the same rates as they arrived during the Hypsithermal.[11] This suggests that there may be some inertia in the response of communities to climate. Part of this inertia occurs because the dominant trees resist invasion of new species by preempting light as well as soil moisture and nutrients. Prior possession appears to confer ownership even when conditions deteriorate, at least for a while. This asymmetry in the sequence of community assembly during warming and cooling periods is known generally as hysteresis. Hysteresis is common in ecosystems with strong positive and negative feedbacks between species and soil resources, but we do not yet have a full understanding of the mechanisms behind hysteresis in ecosystems in response to climate change. This is a question to which **palynology** might contribute key datasets and analyses of the patterns of hysteresis and the different rates at which communities change as the climate warms and cools.

The dry and warm conditions during the Hypsithermal may also have promoted pine forests through higher frequencies of fire. White and red pines have thicker bark that insulates the living cambium against all but the most severe fires. Jack pine's serotinous cones even need fire to melt the resin and open the cone up to release the seeds. Although pines may have had a competitive advantage on sandy outwash because of their high drought tolerance, fires that swept across the outwash plains even once per generation also killed any competitors with thin bark such as maple, birch, and beech, thus reinforcing the occupancy of pines on sands. The hypothesis of the importance of fire in the assembly of the North Woods was tested independently by Albert Swain and Gordon Whitney, who each examined charcoal and ash layers in pollen cores that document past fire frequency.[12] They each found that charcoal and ash layers with pine pollen are common in lake sediments of Hypsithermal age, especially in lakes in landscapes of drier sands than of tills with moister soils.

New large-scale compilations of data such as Neotoma[13] and research consortia such as PalEON (Paleoecological Observing Network)[14] provide exciting new ways to test these and other hypotheses. These large datasets and consortia now allow us to use sophisticated statistical techniques to merge the data from the many pollen cores collected over the years with simulation models to test new hypotheses about rates and patterns of species migrations. These Big Data approaches may indeed provide new answers to how the North Woods assembled itself, answers that have sometimes remained frustratingly just beyond our grasp in the past.

But as Jacqueline Gill points out in her blog "The Contemplative Mammoth,"[15] with this rise in Big Data we may, paradoxically, be training fewer students in the sampling and painstaking analysis of pollen cores. It is easier to compile a record that will impress tenure committees, she says, by writing papers analyzing the older data from these big datasets than to generate new pollen profiles from individual cores, each of which may take a year or more to describe. Consequently, there are fewer courses in pollen identification and pollen analysis, even in former palynological powerhouses such as my own university. Margaret Davis, Herb Wright, and Ed Cushing at the University of Minnesota taught palynology to many colleagues of my generation, some of whose articles I have cited here. When Margaret, Herb, and Ed retired a number of years ago, the University of Minnesota did not hire younger palynologists to replace them. PalEON offers a summer course, not in pollen identification and the preparation of cores for analysis, but in the statistical tools needed for analyzing large datasets of previously published pollen cores. Such new courses are valuable, of course, and no one would argue that we should not offer them. But the emphasis now is on analyzing compiled data rather than obtaining new data in the field.

Does it matter? Yes; courses in pollen identification and analysis are the natural history core of palynology, and we still need such courses. We still need to make detailed analysis of pollen cores from lakes that are carefully chosen to test particular hypotheses about the mechanisms of assembly of the North Woods and other biomes. For example, no one to my knowledge has studied how forests assembled themselves on different soil types in New England to compare with the patterns Linda Brubaker found in Michigan.

A problem with the analyses of large compiled datasets is that the lakes that produced the pollen profiles in them were chosen by past researchers to fill in holes in geographic coverage or to test other hypotheses than those being considered by today's researchers. These compiled datasets therefore are not random samples across the landscape and may not be the best lakes to test today's particular hypotheses. In addition, we now have new laboratory techniques that were not available to previous researchers but could be used together with new pollen cores to test new hypotheses about the assembly of the entire food web and ecosystem. Such techniques include stable isotope analyses,[16] DNA amplification, and methods to identify fungal spores from dung of large mammalian herbivores.[17] As Jacqueline Gill points out, Big Data needs to be fed by Little Data from individual cores. There is still much work to do, hunched over a microscope, squinting at pollen grains.

One of the unanswered questions of the assembly of the North Woods is whether the vegetation was in equilibrium or disequilibrium with the changing climate during deglaciation. A plant community in equilibrium with climate will have a stable species composition over many generations, but a community in disequilibrium with climate will still be changing at an appreciable rate even when the climate is changing slowly or not at all. Answering this question will take analyses of large datasets, modeling, and analyses of new pollen cores from well-chosen lakes. There are a bewildering variety of patterns, unique communities without any modern analogues, and other anomalies that suggest that equilibrium communities, if they exist at all, may be fleeting.[18] Hysteresis of responses to the direction of climate change, time lags in range expansions caused by slow dispersal of seeds and other life history strategies, a crisscross pattern of species range expansions, and the local effects of glacial landforms all complicated the assembly of the North Woods. Margaret Davis argues that these complications ensure that the North Woods was never in equilibrium with climate.[19] Herb Wright disagrees, concluding that vegetation responds comparatively quickly to climate change, within a generation or two,[20] so equilibrium communities may have persisted for several generations or more. Equilibrium gives the plant species that occupy a site more time to influence the development of the soils through the decay of hundreds of annual cohorts of litter. Long-term site occupancy by a community at equilibrium with climate also allows the development of a more complex food web as populations of different animal species become established on a site only after their preferred habitat is sufficiently stable for long periods.

Part of the difficulty of determining whether plant communities were at equilibrium with climate lies in the spatial scales across which the plant communities are characterized. Beneath the crown of a large tree, vegetation is never in equilibrium: The tree grows, younger trees grow under it, the dominant tree dies, and the young trees compete for the light in the gap. At this scale, vegetation is always changing and never in equilibrium. But across the landscape, the dynamics of the life cycles of individual trees average out. Does vegetation come to an equilibrium with climate across the larger landscape, and if so, at what scale? Can we determine this from the pollen record? Fortunately, different-sized ponds and lakes sample the landscape at different distances. The pollen flux into large lakes is dominated by regional pollen from large distances; pollen flux into small ponds is dominated by nearby local pollen sources.[21] Therefore, lakes and ponds are lenses that scan the landscape with different focal lengths that depend on their radii. It should be possible to determine whether equilibrium or disequilibrium in the pollen record depends on spatial scale by sampling ponds and lakes of different sizes but in the same outwash plain or moraine. However, I know of no studies that have systematically taken on this research program, which will almost certainly require descriptions of new pollen cores.

Another open problem in the pollen record concerns the origins of many anomalous communities with no modern analogue. Tom Webb suggests that these anomalous communities may arise from the individual and unique responses of different species to climate.[22] Webb notes that different species respond not to "climate" but to different properties of the climate. For example, he showed that beech expanded northward into Quebec from 6,000 to 4,000 years ago while spruce expanded southward. This beech–spruce community has no modern analogue. Such opposite directions of beech and spruce migration in the same area at first seem contradictory: The southward movement of spruce suggests a general cooling, but the northward response of beech suggests overall warming. But Webb suggests that calculations of the orientation of the earth relative to the sun at that time indicate that summers were cooler but winters were warmer then, on average, than periods before and after it. The beech may have moved northward because of warmer winters, which beech could survive without frost-cracking. At the same time, cooler summers may have allowed spruce to survive without heat stress. Anomalous communities may also form with future global warming, and the study of such communities in the pollen record may help us anticipate when and where we may begin to see them in the coming decades.

These different theories of vegetation response to climate warming since deglaciation all depend on the underlying assumption that the climate changes independently of vegetation and, indeed, that the changes in vegetation are being forced by the changing climate. However, this is not the only way to think about this question. What if we consider the vegetation and climate to be two components of a larger system? Changes in both vegetation and climate might result from feedbacks between the two. Vegetation, snow, and ice all affect the energy balance of the landscape and the entire globe because dark vegetation absorbs sunlight, whereas snow and ice are almost perfect reflectors of it. As the dark conifers invaded the previously blindingly white landscape, they themselves may have absorbed a larger portion of the sun's heat, which may have warmed the climate and melted the ice sheet further.[23] A merger of large pollen datasets, new pollen cores, and simulation models such as those being undertaken by PalEON might help shed light on this issue. These new models and statistical techniques will allow us to make rapid progress, but the bedrock of the science remains the gathering of basic natural history descriptions of species movements from detailed descriptions of pollen cores.

Meanwhile, this spring's pollen is settling to the bottoms of lakes everywhere, providing a record of today's surrounding forest; this pollen will be buried beneath other layers of pollen to be produced by generations of trees yet to come. Palynologists far into the future might study these pollen records to learn how the North Woods we know today responded to the current warming of the globe.

3.

Beaver Ponds and the Flow of Water in Northern Landscapes

The ice sheet may have sculpted the landscape, but beavers now control the hydrology of much of the northern half of the continent.

After the ice sheet retreated, after the drainage patterns became organized, after the North Woods plant communities assembled themselves, herbivores such as moose, deer, and caribou quickly followed. The arrival of these herbivores signaled the beginnings of northern food webs. Many of these herbivores in turn controlled the distribution and growth of the plants they ate and therefore the composition of the plant communities. But no herbivore in the North Woods, and few anywhere on Earth, had as large an effect on the landscape as the beaver.

Beavers are the animal that drew the early explorers and naturalists to the North Woods as they searched and traded for furs to satisfy the demands of the European fashion scene. Except perhaps for whales, there may be no other single animal that is responsible for such an extensive exploration and exploitation of any biome. Although the beaver's original range stretched from the Arctic to the Gulf of Mexico, its population density and the quality of its fur were highest from the Great Lakes northward.

The flow of water from the northern half of this continent to the sea is controlled today as much by the beavers' dams and ponds as it is by how the great Laurentide Ice Sheet sculpted the land surface. Today, beaver ponds and the wet meadows that form after beavers abandon them occupy 15 percent of all the land area of northern Minnesota.[1] In the wilderness areas of northern Minnesota, where trapping is not allowed and beaver populations have recovered to pre–fur trade levels, more than 90 percent of the water that drains into the lakes flows through at least one beaver pond. That's many millions, even billions, of liters of water per second that are controlled by this little rodent.

Beavers prefer to build dams in places where a stream flows through a constricting gap in bedrock ridges or, more usually, in the end moraines sweeping across the landscape. Usually, the dam plugging the gap floods a broad, shallow basin upstream that was hollowed out by the ice sheet. A beaver is attracted to these gaps because of the gurgling sound of water flowing through them. Beavers therefore appear to search for particular basin geometries suited to establishing their ponds. In a sense, the beaver "hears" the geometry of the basin in the music of the bubbling stream.

The size and shape of the basin amplifies the changes beavers make to the landscape when they build these dams. The oldest and usually the largest ponds are created where a small investment in building a dam forms a large pond. These large ponds can sometimes provide access to a sustainable supply of aspen and willow, the beavers' preferred food, growing in the pond's riparian zone and in the uplands ramping up from it. After the aspens and willows are cut, new shoots will sprout up from their roots unless they are shaded by understory spruce and fir. These young aspens and willows subsequently grow to the larger sizes preferred by beavers. By harvesting aspen in different places each year around large ponds, beavers often manage to achieve a sustained yield between their annual harvests and the annual regeneration of young aspens and willows. Therefore, these large ponds can often be occupied almost indefinitely; they form the nucleus for the dispersal of the younger generations to other areas up and down the same valley or to other valleys.

Parents aggressively displace their young kits from their home by the time the kits enter their second year. If the kits were allowed to stay, the growing family would overrun its food supply, resulting in severe mortality in the population. These young beavers must find their own places to build dams and make ponds. Because earlier generations of beavers have claimed the best locations for ponds, its newer members must make do with more marginal places. These marginal ponds tend to be in smaller basins. The smaller sizes of these ponds provide the beavers less access to food than the older and larger ponds. Aspen and willow around smaller ponds are often depleted before they can regrow to the large sizes preferred by beaver. The smaller, marginal ponds are therefore more transient and are occupied for shorter periods of time than the larger ponds in more preferred areas.

Within several decades, generations of beavers residing in an area will completely terrace the valleys with ponds and wet meadows, each pond or meadow rising above the previous one in steps determined by the height of its dam. (One of my wife's uncles was fond of saying that life is like a beaver colony, one dam thing after another.) Taken together, ponds and meadows form a giant stairway snaking up the valley. Beaver ponds are in many ways the organizing force of the hydrology of these valleys.

A beaver dam is a fascinating structure, not a random pile of sticks as you might expect. I have become a real connoisseur of dams, having seen some truly wonderful landscape sculptures that the artist Christo would envy. I was fortunate once to see in Alaska the first layer of sticks laid down by a beaver colony building a new dam. Each stick was actually a branch with several smaller branches radiating from one end. The main stems were aligned parallel to one another in the direction of stream flow, with the smaller radiating branches facing upstream. Sediment was being trapped in the forks of the smaller branches, anchoring them firmly to the stream bottom. This first row of branches was carefully placed to form a solid foundation atop which the rest of the dam could be built.

Above this foundation, the rest of the dam is composed of sticks, mud, and stones, carefully placed to hold back the rising mass of water behind it. Many dams are horseshoe-shaped and convex upstream, much like Hoover Dam. This upstream convexity locks the dam into place as the force of water builds against its upstream side. As sediment builds up against the upstream face, it further strengthens the dam against the pressure of the current. The upstream slope of the dam gradually becomes less steep, but the downstream side of the dam is steeper and remains so as water flows over it. To protect the downstream side against erosion by the flowing water, beavers often face it with sticks 8 to 16 centimeters in diameter and a meter or two long, carefully placed vertically on the downstream face. Beavers do not waste aspen and willow, their preferred foods, in building the dams. Instead, their preferred building material is alder,[2] which is as abundant as aspen and willow along the banksides and as easy to cut but shunned as food because of its astringent taste from phenolic compounds in the bark.

As the height of the dam grows, often to 2 meters or more, the pond behind it deepens. It is in this deep pool immediately behind the dam where the lodge is often located. Most of the mass of the lodge is below water, much like an iceberg. It is always a surprise to see how huge the lodge really is once the beavers abandon the pond and it drains when the dam is no longer maintained.

While the deepening water protects the beavers and their lodges from predators, it also threatens to burst the dam because of increasing pressure at its base. To counter this pressure, beavers often build secondary

dams a few meters downstream from the main dam. These secondary dams create smaller ponds backing up water against the main dam and thereby equalize the pressure from upstream at the dam's base. One way to test this hypothesis is to determine whether the presence of secondary dams correlates with the height of the main dam. The higher the main dam, the deeper the water behind it, the greater the water pressure at its base, and the greater the need for secondary dams to stabilize the main dam. I have not yet done this test, however.

Curiously, the secondary dams often are not as well constructed as the main dam. Perhaps the beavers have learned not to spend too much energy and time on these dams, because they are merely insurance policies against failure of the main dam. But I have always thought that these dams are also where parents teach their young the arts of dam construction. This would explain the obvious mistakes in many of these secondary dams, such as weaker stretches of solid mud without the reinforcement of sticks, sloppy placement of the downstream-facing sticks, and a crooked line instead of an upstream convex shape. This hypothesis could be tested using trail cameras placed next to secondary dams. If both parents and young come out at night to repair these dams together, then this would give support to the hypothesis of a teaching function for these dams.

As they age, beaver dams progress through four stages: (1) overflow dams, which are generally intact structures maintained by an active colony with the water flowing over their tops; (2) gap dams, with water flowing through a small gap on the top edge of the dam, sometimes intact and maintained by an active colony but sometimes also in the first stage of degeneration after abandonment; (3) underflow dams, with water leaking through holes in the bottom, which are generally found in abandoned ponds (I suspect that these holes are often made by otters or muskrats tunneling through the dam itself); and (4) throughflow dams, which leak water throughout their downstream face, generally unmaintained and abandoned and forming the final stage in the decay of the dam, and which are often ready to blow out.[3] The sequence of the dam moving through these four categories can be viewed as the natural process of development or aging of the dam. Biologists call the process of development of organisms throughout their lifespan their ontogeny, so we can think of the progression of the dam through these four stages as the ontogeny of the dam.

The wetlands behind these dams also follow their own ontogeny as the dams age.[4] This ontogeny of wetlands in beaver ponds and meadows, driven by the ontogeny of the dams, is responsible for the very high plant species diversity in valley bottoms occupied by beaver populations. Overflow and gap dams have large ponds of deep water behind them, with floating lily pads in the water and emergent cattails and northern blue flag irises along the water's edge. The breaching of the dams by leaks at the base (underflow) or throughout the entire structure (throughflow) lowers the water table and allows a different type of wetland to form, a wet meadow that progressively dries as the pond drains over several years. These meadows are occupied by a great diversity of plant species,[5] mainly grasses and sedges but also Canada anemones, wild geraniums, thistles, goldenrods, asters, violets, ferns, and many other species. There are often more than sixty plant species in these meadows, much more than in a similar area in the surrounding upland aspen–birch–spruce–fir forests. Not only is the species diversity of the wet meadows very high, but the productivity of the meadow plant community is sometimes even greater than that of the surrounding forest.

Ground and surface water flows through these meadows[6] in complex pathways that we are just beginning to understand, bringing with it the essential nutrients that sustain this productivity and diversity. Sediment is trapped and soil organic matter accumulates during this succession of plant communities to levels three or four times what was present in the same square meter of the stream that preceded the pond. Eventually, sometimes after many decades, willows, aspen, and alder invade the meadows and shade out the grasses and sedges. As their food supply of aspens and willows grows, the beavers return, repair their dam, flood the valley, and restart the life cycle of the pond. The new pond buries the older one under a new layer of sediment. It is entirely possible that the dams and sediments of several beaver ponds lie stacked atop one another in the basin behind the dam, like layers of a cake in a bakery.[7]

The geometry of the basins carved by the ice sheet is the stage for the building of the dam and the pond behind it, but the building and decaying of the dam determine the dynamics of the pond and meadow and the flow of water down the valley. The great diversity of wetlands in these meadows derives from this continuous interplay of the ontogeny of the dam with the geometry of the basin. If the geology of the basin provided the stage, then the waxing and waning of the beaver populations, their dams, and their associated plant communities make up an ecological play in several acts.

4.

David Thompson, the Fur Trade, and the Discovery of the Natural History of the North Woods

David Thompson, a trader and an explorer in the beaver fur trade, compiled some of the most extensive observations of the natural history of the North Woods and the boreal forest in the decades after the American Revolution.

On May 21, 1797, David Thompson quit the Hudson's Bay Company and began walking 80 miles to the Nor'West Company's trading post at Reindeer Lake in northern Manitoba. Upon arrival, he immediately embarked on a 1,600-kilometer inland voyage in a 6-meter-long birch-bark canoe with a brigade of *hivernants*, French voyageurs originally from the Loire Valley. The *hivernants* wintered over in the Far North and traded with the Chippewayan Indians, the central and largest of all the Athabascan tribes, for beaver, lynx, snowshoe hare, and other furs. His route took him southeast down the chain of kettle lakes inside the terminal moraine to Lake Winnipeg, up the Winnipeg River to Lake of the Woods, then up the Rainy River to Rainy Lake. From here, he proceeded east through the North Woods of present-day Minnesota and Ontario across numerous lakes carved by the ice sheet. On July 22, Thompson, the *hivernants*, and the cargo of furs arrived at the western end of the Grand Portage at the head of the Pigeon River flowing into Lake Superior.

From their high vantage point at the western end of the Grand Portage, the voyageurs portaged close to 1,400 kilograms of furs from each canoe over the 14-kilometer trail next to the Pigeon River down to the trading post at Hat Point, on the shore of Lake Superior. At this large and busy trading post, the furs were loaded into giant 13-meter freighter canoes with a capacity of 4 tons. The freighter canoes, each paddled by ten voyageurs known as *mangeurs de lard* (pork eaters), were then paddled another 2,000 kilometers across Lake Superior, then Lake Huron, Lake Erie, Lake Ontario, and then down the St. Lawrence River to Montreal, the headquarters of the Nor'West Company and Thompson's new employer. This route took them through the heart and across almost the full breadth of the North Woods. At Montreal, the furs were loaded on ships bound for Europe, where the gentlemen of London wore the beaver furs made into top hats and the ladies wore the lynx and hare furs made into muffs and coat collars.

Hudson's Bay Company and the Nor'West Company had different business models of fur trading, and the Nor'Westerners' philosophy suited Thompson much better. The Hudson's Bay Company was content mainly to establish posts around the shore of Hudson's Bay and let the Indian trappers come to them with furs. They had no interest whatsoever in exploring the boreal forests stretching westward. Joseph Robson, a surveyor hired by Hudson's Bay Company, wrote, "The Company have for eighty years slept at the shores of a frozen sea; they have shown no curiosity to penetrate further themselves, and have exerted all their art and power to crush that spirit in others."[1]

Thompson, who also had been trained in Britain as a surveyor, discovered within himself a boundless curiosity in nature and Native Americans after arriving at the Hudson's Bay post at Churchill in 1784. During his employment, he was warned several times by the Hudson's Bay governors to end this exploration nonsense and stick to making money (for them, I might add). Apparently, in 1797 he felt his spirit crushed enough to switch allegiances to the Hudson's Bay Company's rivals.

The Nor'West Company, in contrast, welcomed Thompson's skills in surveying and natural history. Montreal is 1,600 kilometers farther away from the land of prime beaver pelts than the Hudson's Bay Company's posts. To compete with the Hudson's Bay Company, the Nor'Westerners had to employ voyageurs to skirt around the southern and western flanks of the Hudson's Bay Company's empire. The lakes, both great and small, carved by the great ice sheet 10,000 years ago along the edge of the granitic Canadian Shield, formed a wilderness highway from Montreal to the Arctic over which the canoes and their cargo could be conveyed. The Nor'West Company needed someone who could survey this route and make accurate maps and observations of the natural history and native customs of this vast land, known as *le pays d'en haut*, or the Upper Land. Thompson was more than willing to oblige.

And succeed he did. In the coming years Thompson traversed this route through the North Woods up the Athabascan River to its headwaters in the Canadian Rockies. Here, in a bewildering array of confusing mountainous drainages, he discovered the nearby headwaters of both the Fraser and the Columbia Rivers and then traveled down the Columbia to its mouth in 1811, the first European to travel the full length of this great river, 6 years after Lewis and Clark's expedition.

During his travels, Thompson painstakingly recorded his observations of nature, the customs of the Indians, and the latitudes and longitudes of key locations. Although John and William Bartram and Peter Kalm had begun to lay the foundations of the natural history of the North Woods a few decades before Thompson arrived at Hudson's Bay, none of these gentlemen exceeded Thompson in the geographic breadth and ecological and anthropological depth of their writings. Thompson kept field journals and notebooks during his entire time in North America, which he only later began compiling into his *Travels* for publication. However, he went blind in 1850 before completing this book. He died shortly after, in 1857. This unfinished manuscript was rediscovered, along with Thompson's field notes, leatherbound journals, and maps, by Canadian geologist and arctic explorer J. B. Tyrell, who edited and published them in 1916 in a limited edition of 500 copies, two of which reside in the Northeast Minnesota Historical Archives in the library of my campus. This version has been recently reissued[2] jointly by the Champlain Society, the McGill–Queen's University Press, and the University of Washington Press, with superb editing and commentary by William Moreau.

Thompson always saw his field notes and journals as a source of valuable data and observations on the northern environment for his scientific colleagues in North America and Europe. This golden age of natural history in the eighteenth and nineteenth centuries was a period of intense theorizing about the age and origin of the earth and the classification of living things, culminating in Darwin's *Origin of Species*. A famous example of Thompson's exquisite observations and writing is his description of the mosquito's proboscis:

> The Mosquitoe Bill, when viewed through a good Microscope [a good microscope in a canoe somewhere in the northern wilderness!], is of a curious formation; composed of two distinct pieces; the upper is three-sided, of a black color, and sharp pointed, under which is a round white tube, like clear glass, the mouth inverted inwards; with the upper part the skin is perforated, it is then drawn back, and the clear tube applied to the wound, and the blood sucked through it into the body, till it is full; thus their bite are two distinct operations, but so quickly done as to feel only one.[3]

Nothing escaped his curiosity. He described a polar bear feasting on beluga whale on the shorefast ice of Hudson's Bay and a black bear fishing for trout; the habitat and growth of spruce, fir, aspen, birch, sugar maple, and wild rice; the behaviors of loons, mergansers, ravens and crows, bitterns, snow buntings, sharptailed and spruce grouse, crossbills, cormorants, boreal chickadees, Canada and snow geese, sandhill and whooping cranes, trumpeter and tundra swans, goshawks and gyr-falcons, sharp-shinned hawks and peregrine falcons, both golden and bald eagles, ospreys, and boreal, snowy, great grey, and northern hawk owls ("the meat of the Owls is good and well tasted to hunters"[4]); how to catch pike, whitefish, carp, lake and brook trout; and the habitats and behaviors of arctic and red fox, caribou (which he calls Rein Deer), moose, bison, martens, ermines, lynx, wolves and wolverines, minks, and muskrats. He recorded maximum and minimum temperatures in various locations, described how chinook winds warm the land imme-diately east of the Rockies, and explained in detail the appearance of the aurora borealis and its effect on compass needles.

And, of course, he was fascinated by the ecology of the beaver. Here is part of his description of beaver population dynamics and how their ponds and dams control the hydrology of the North:

The Beaver were safe from every animal but Man [by which he means the native Indians], and the Wolverene. Every year each pair having from five to seven young, which they carefully reared, they became innumerable, and except the Great Lakes, the waves of which are too turbulent, occupied all the waters of the north-ern part of the Continent. . . . To every small Lake, and all the Ponds they builded Dams, and enlarged and deepened them to the height of the dams. Even to grounds occasionally overflowed, by heavy rains, they also made dams and made them permanent Ponds, and as they heightened the dams increased the depth and added to the extent of the water; Thus all the lowlands were in possession of the Beaver, and the hollows of the higher ground. Small streams were dammed across and Ponds formed; the dry land with the dominions of Man contracted, everywhere he was hemmed in by water without the power of preventing it; he could not diminish their numbers half so fast as they multiplied; and their houses were proof against his pointed stakes, and his arrows could seldom pierce their skins. In this state, Man and the Beaver had been for many centuries.[5]

Besides making keen observations, Thompson executed experiments, without controls and unreplicated to be sure but experiments nonetheless, to test hypotheses about animal physiology and behavior. Upon killing a caribou, he was surprised at how unusually warm the blood felt. Opening its stomach, he found it was

full of a white moss [which William Moreau thinks is the reindeer lichen *Cladina rangiferina*]. I tasted it and . . . it was warm to my stomach. I then traced the Deer to where they had been feeding, it was a white crisp moss in circular form, of about ten inches diameter each division distinct, yet close together. I took a small piece about the size of a nutmeg, chewed it, it had a mild taste, I swallowed it, and it became like a coal of fire in my stomach, I took care never to repeat the experiment.[6]

On another occasion, he noticed that owls seem not to eat captured mice until the owl is sure that the mouse is dead. He hypothesized that the owl decides that the mouse is dead when it no longer moves. He tested this hypothesis by poking a dead mouse with a willow twig in front of a tame owl, whereupon the mouse "instantly received another crush [with the owl's] beak and thus continued until it was weary, when loosening its claws it seized the Mouse by the head [and] crushed it. . . . I concluded, that to carnivorous birds, the death of its prey is known by the cessation of motion."[7]

One of the unsolved mysteries of Thompson's journals is the identity and existence of what the Chippewayan Indians called *Mah thee Mooswak* (the ugly moose). Thompson notes,

It is found only on a small extent of country mostly around the Hatchet Lake. . . . This deer seems to be a link between the Moose and the Rein Deer; it is about twice the weight of the latter; and has the habits of the former; it's horns are palmated somewhat like those of a Moose, and its colour is much the same; it feeds on buds and the tender branches of Willows and Aspins, and also on moss. In all my wanderings I have seen only two alive.[8]

William Moreau thinks this is probably the woodland caribou,[9] but although Thompson's description of the animal's diet fits that of the woodland caribou, his description of its appearance does not. Woodland caribou are at most only a few kilograms larger than the reindeer, or barren-ground caribou, with which Thompson was intimately familiar. Both woodland and barren-ground caribou are instantly recognized as being related, and taxonomists today classify them as two subspecies of *Rangifer tarandus*. Although it does not form the vast herds of barren-ground caribou, the woodland caribou is not uncommon and is frequently seen in small groups, so it is unlikely that "in all his travels" Thompson would have seen only two isolated individuals. Moose today consists of at least seven subspecies within one species (*Alces alces*). Although modern molecular studies of the evolution and diversification of moose are just beginning,[10] the evidence points to a radiation into these subspecies when the North Woods was becoming assembled at the end of the Wisconsinan Glaciation about 10,000 years ago. It may be possible that Thompson saw the last vestige of an isolated subspecies of moose that may have since gone extinct.

Thompson was as much an anthropologist as he was a surveyor and ecologist. He wrote about how Eskimos on the shore of Hudson's Bay build their igloos and paddle their kayaks; about the Chippewayans' fishing and hunting techniques, their social structures, their religious beliefs and origin stories; and about the relationships between the various tribes of the Canadian prairies. He compiled some of the first vocabulary lists for several Indian languages. Thompson wrote about the Indian cultures without an overlay of European bias and judgment, more than 100 years before Franz Boas and Ruth Benedict made objective and unbiased field studies the acceptable method in cultural anthropology.

Despite his being a loyal employee of the Nor'West Company, Thompson foresaw what iron tools and traps provided by the fur trade were doing to both the Indians and the beaver. In a continuation of his description of the ecology of the beaver quoted earlier, Thompson wistfully writes,

Without Iron, Man [meaning the native Indians] is weak, very weak, but armed with Iron, he becomes the Lord of the Earth, no other metal can take its place. For the furrs which the Natives traded, they procured . . . Axes, Chissels, Knives, Spears, and other articles of iron. . . . Thus armed the houses of the Beaver were pierced through, the Dams cut through, and the water of the Ponds lowered, or wholly run off, and the houses of the Beaver and their Burrows laid dry, by which means they have become an easy prey to the Hunter.[11]

One senses from this and other passages that Thompson wished he simply could have explored the North Woods, which he so obviously loved, without the need for annihilation of animal populations or human cultures. It is only in the past four or five decades that the beaver populations of the North Woods have recovered from the trapping that Thompson observed and participated in. Reading Thompson's journals today, any ecologist would wish to be with him in his canoe, learning from him but at the same time knowing how it would all turn out. Perhaps things would have turned out differently if that could have been done.

Postscript: As I write this, the tar sands along the Athabasca River are being mined, spewing toxic pollution into the river in which David Thompson's canoe floated while he wrote about the beauty of this land. Oil and gasoline will be refined from these tar sands; when they are burned, carbon dioxide will be emitted into the atmosphere, where it will trap heat and alter the climate that was responsible for the assembly of the North Woods through which Thompson traveled.

PART II
Capturing the Light

The North Woods, like almost all of Earth's ecosystems and food webs, is driven by the capture of light through plant pigments, located mainly in leaves, and its storage in the chemical bonds of sugars and other carbohydrates. Although the species diversity of tropical forests easily eclipses that of the North Woods, there is far less diversity in the tropics in the lifetimes of leaves, their shapes, and the shapes of the crowns of tropical trees. No other biome matches the North Woods for the diversity of leaf lifespans, the diversity of shapes of leaf surfaces, and the diversity of crowns of different shapes, from the cone-shaped crowns of conifers to the rounded crowns of deciduous hardwoods.

A leaf is not simply a flat surface acting as a solar panel. No matter how thin or skinny, leaves have a volume occupied by three or four layers of cells that contain the machinery of photosynthesis. The outermost layer, the **epidermis**, is composed of cells with waxy or resinous cuticles to prevent water loss. Light passes through the epidermal layers and triggers photosynthesis in the cells of the inner layers, known as **mesophyll**. Carbon dioxide enters the mesophyll through pores in the epidermal layer, known as **stomates**. Water and nutrients are brought up from the soil to the leaves by a network of pipes called **xylem**. So, thin as leaves are, a lot of anatomical structure is packed into their volume for the machinery of photosynthesis. The shape of the leaf is a geometric solution to the problem of presenting area to the sun to capture light and funnel it down to the mesophyll and capture carbon dioxide through open stomates while simultaneously minimizing water loss. The tree must arrange its leaves throughout its crown to maximize the capture of light and carbon dioxide and minimize the loss of water to maximize photosynthesis across their lifetimes. The lifetime and shape of the leaf, the shape of the crown, and the arrangement of leaves within it are solutions to this problem. Natural selection and evolution are nature's ways of finding these solutions for particular environments and passing them down to offspring. How natural selection has found a variety of solutions for species in the North Woods and what they mean for their coexistence are the subjects of this section.

5.

How Long Should a Leaf Live?

The evolutionary and economic tradeoffs of leaf lifespans of deciduous and evergreen trees.

One day in spring, buds that have withstood the northern winter begin breaking and sending forth fresh new leaves of an astonishing variety of green hues. As they unfold from the buds, these young leaves are smaller than a squirrel's ear, but in the next few weeks they will grow rapidly into little solar panels and begin to capture sunlight. Deciduous leaves of maple, oak, aspen, birch, and other species will live only for a few months before their green color gives way to the saturated reds, yellows, and oranges of autumn. The needles of pines will live for 2 or, less commonly, 3 years before they are shed. Spruce and eastern hemlock needles can live for at least 4 and sometimes 10 or more years. There are few ecosystems anywhere with such a wide range of leaf lifespans as the North Woods.

The lifespan of a tree's leaves is an important part of the tree's life cycle: The leaf's lifespan determines how much sugar it will produce; how long it is exposed to the wind and weather that batter it and the insects, snowshoe hare, moose, and deer that eat it; and how long the tree must invest nutrients in its maintenance. Supporting and maintain-

ing a leaf throughout its lifespan therefore poses a variety of challenges to a tree. Why do some species retain their leaves through several winters whereas others shed them every year? How does the lifespan of a leaf relate to its photosynthetic rate and its chemistry? Why does the North Woods have such a diversity of leaf lifespans?

To understand why leaves have different lifespans, think of them as investments a tree makes that have benefits and costs.[1] This is an economist's view of a leaf. The major benefit leaves provide to trees is the capture of light energy by green chlorophyll, which is then used by photosynthesis to convert carbon dioxide from the atmosphere and water from the soil into sugars. The sugars are then made into wood, bark, and roots. The total costs of a leaf, on the other hand, are all the investments a tree must make to construct the leaf and then maintain the machinery of photosynthesis. Given all this, how long should a leaf live to maximize benefits while minimizing costs?

We will simplify the problem for the moment by focusing only on the gains, losses, and construction requirements of capturing carbon to produce sugars and other compounds that become the tree's biomass, setting aside for now the additional costs of capturing water and nutrients. Carbon is the most abundant element in a leaf, making up about half its dry weight. The objective of producing a leaf is to maximize carbon gain per unit time by photosynthesis while paying the carbon costs of constructing and maintaining the leaf structure. The costs of making a leaf are analogous to the costs of building a factory. The maintenance costs of manufacturing sugars from sunlight, carbon dioxide, and water are paid as long as photosynthesis continues, much as the costs of electricity and maintenance staff are paid out as long as the factory is manufacturing something. In all leaves, the rate of photosynthesis slows as the leaf ages, and so maintenance costs also decrease.

In order to calculate the leaf lifespan that maximizes net daily carbon gain while balancing construction and maintenance costs, we need a model that can keep track of all comings and goings of carbon. The model is an accounting of the carbon budget of a leaf, involving many hundreds of reactions and compounds. But a model with hundreds of equations (one for each reaction or compound) does not help us reduce the complexity of a leaf to simpler terms so that we can understand it. The objective of model building is to present a reasonable simplification that includes the important details of a particular process, not to be an accurate mirror of all the details of nature. The best way to construct a model is to start as simple as possible (but, in Einstein's wise words, no simpler) and add complications only when needed.

The art of model building consists of zeroing in on the most important details first and ignoring all the rest. For a model to provide insight into biological problems, every term and parameter must have biological meaning. The important details and biological meanings of the model's terms and parameters reside in the natural history of the organism. A sound knowledge of natural history is therefore essential to developing the mathematics of the proposed model. Failure to pay attention to natural history while constructing a model may lead to mathematically elegant but biologically meaningless conclusions.

Perhaps the simplest model of carbon gains and losses and leaf life-spans was constructed by Kihachiro Kikuzawa of the Hokkaido Forest Experiment Station in Japan's equivalent of the North Woods.[2] Kikuzawa's model was one equation that included three terms for net carbon gain per unit time: one term for the gain of carbon by photosynthesis and two terms for the allocation of carbon for maintenance and for construction. He assumed that photosynthesis and maintenance rates decline as the leaf ages. The leaf begins life by paying its construction costs up front, so it starts out with a carbon deficit. This initial deficit is like a mortgage taken up front to build a factory; it is then paid back by photosynthesis during the leaf's lifespan. Finally, Kikuzawa made one more simplifying assumption: The plant produces one leaf at a time. This last assumption made the problem simple, because all Kikuzawa had to do is calculate when the plant should let this leaf die and grow a new one. Admittedly, there is no plant anywhere that works exactly like this, but this simple model captures the important natural history of construction, maintenance, and photosynthesis common to all leaves.

In Kikuzawa's model, as photosynthesis incorporates more carbon dioxide into growth, the leaf begins to show a carbon profit over the initial construction investments. As the leaf ages further, the rate of photosynthesis slows, and so does the rate of increase in the total net carbon gained over the life of the leaf. But the leaf does not simply maximize the total net carbon gain; if that were the case, the leaf should continue to live as long as net carbon gain increases. Instead, the leaf maximizes the average net daily carbon gain, which is the total net gain (photosynthesis minus maintenance and construction) up to that day divided by the number of days the leaf has been alive. The average daily gain at first rises when the leaf is young because photosynthesis outpaces maintenance costs in the numerator of this ratio, and the number of days the leaf has been alive in the denominator is small. At some point, the average net daily carbon gain reaches a maximum and then declines because photosynthesis in the numerator slows, but leaf age in the denominator continues to rise. The age of the leaf at this maximum is the optimal age a leaf should live. Living shorter than that is not a good strategy because the leaf can always do better by living another day (the average net daily carbon gain is still increasing), but living longer than that is also not a good strategy because the average net daily carbon gain is now dropping.

Kikuzawa's model predicted that leaf lifespan should be short when photosynthesis declines more rapidly than maintenance as the leaf ages. Leaf lifespan should be long when the costs of constructing a leaf are large, because it will take a long time to pay these costs back. As Kikuzawa drily remarked, "These points do not conflict with the empirically observed facts." Deciduous leaves such as those of maple, birch, and aspen have high photosynthetic rates that drop rapidly below the maintenance costs as the season progresses, whereas evergreen needles from spruce, pine, and fir have high construction costs and low but steady photosynthetic rates that stay above maintenance costs for much longer. During the height of summer, deciduous leaves may photosynthesize at two to four times the rates of evergreen needles. On the other hand, photosynthesis can be so slow in some evergreen needles, even under the best conditions, that the needle must remain alive for several years to recover the initial construction investment.

But there are additional costs of water loss and acquisition of nitrogen for the photosynthetic machinery. In order to get carbon dioxide to its photosynthetic machinery, the leaf must open its stomates, which are small pores on its surface. A leaf cannot actively pump carbon dioxide into its stomates; it relies instead on the carbon dioxide diffusing passively through the stomatal pores. High rates of photosynthesis require that stomates be kept open longer to obtain the needed carbon dioxide. However, stomates are a two-way gate: As carbon dioxide enters the leaf through the open stomates, some of the water brought up from the roots exits the leaf through them, passing the carbon dioxide molecules on the way. Water is essential to a leaf for three main reasons: to combine with carbon dioxide to make sugars during photosynthesis, to keep its cells from collapsing, and to serve as a medium for transport of sugars, nutrients, proteins, and other essential molecules. So the loss of water through the open stomates is a considerable cost to the leaf. For each carbon dioxide molecule that enters the leaf, several water molecules escape to the atmosphere. The unavoidable loss of water molecules that the plant must pay in order to photosynthesize is known as **transpiration**, and it is a major constraint on the lifespan of a leaf.

Nitrogen is the major nutrient in the photosynthetic machinery because it is the key element of the amino acids in the protein called ribulose-1,5-bisphosphate carboxylase/oxygenase, known more simply as RuBisCO. RuBisCO catalyzes the initial step in the incorporation of carbon dioxide into plant biomass. Because plants constitute 99 percent of all living biomass on Earth and RuBisCO accounts for one third of protein in leaves of most plants, RuBisCO is the most abundant protein on Earth.[3] In full sunlight, the amount of RuBisCO in a leaf limits the rate of carbon dioxide incorporation and therefore plant growth. The more RuBisCO the leaf has, the more biomass it can make, but the more nitrogen it needs for the RuBisCO molecules. Getting enough nitrogen to maintain large amounts of RuBisCO by uptake from the atmosphere through nitrogen fixation or from the soil is therefore a large maintenance cost for the leaf and an additional constraint on the lifespan of a leaf. This is why the supply of nitrogen from the soil into the plant during uptake and eventually into RuBisCO limits productivity of terrestrial ecosystems,[4] including the North Woods.[5]

In the upland forests of the North Woods, the supplies of both water and nitrogen are correlated because the dryness of soils on sandy out-wash plains inhibits microbes from decomposing leaf litter and releasing its nitrogen for subsequent reuptake by the plants, whereas soils on the finer-textured moraines can hold more water and support faster decay and nitrogen release. The cold and permanently water-saturated soils of bogs and fens in the peatlands that occupy former glacial lake basins are an exception to this correlation between water and nitrogen supplies. These wet soils have low rates of nitrogen supply because their permanent saturation keeps them cold and devoid of oxygen, which inhibits decomposition.

Because of the high costs of nitrogen in photosynthetic machinery and the high unavoidable losses of water through the stomates, most deciduous species with high photosynthesis rates pay the price by being able to survive only in places where nutrients, water, or light, preferably all three, are abundant. So we find maples growing mostly on clay soils that contain sufficient water and nutrients and aspens and birches growing in open areas after fires or logging, with full sunlight and high soil nutrient availability.

When times are tough, such as the onset of frost in autumn, deciduous leaves are discarded. Although the trees must construct a new set of leaves next year, the rapid photosynthetic rates of deciduous species during the growing season, made possible by adequate supplies of water and nitrogen, more than compensate for the loss. The deciduous strategy is like an investor who buys stocks at rapidly rising prices but sells the minute the stock price drops. Such investors and plants need large initial investments to make any amount of money or sugars in a short time.

If deciduous hardwoods are the flashy investors of the woodland economy, then evergreen conifers are the more fiscally conservative members of the forest, adopting the policy of holding on to prudent investments with slow but steady returns. Evergreen conifers store nutrients in the needles for 2 or 3 years, creating an internal pantry from which the necessary ingredients for photosynthesis can be supplied when the soil cannot provide them, such as when precipitation is low or the soil is frozen. As a result, conifers can outcompete deciduous species on dry and nutrient-poor soils or in peatlands.[6] Moreover, photosynthesis can continue at lower temperatures in evergreen needles than in deciduous leaves. This allows conifers to capture and store carbon during brief thaws in winter and begin growing earlier in spring, thereby lengthening the growing season. This ability of evergreens to cope with nutrient-poor soils and cold environments may be one reason why they dominate northern forests.

The evergreen needles of northern conifers must be protected for several years against abrasion by snow and ice and the desiccating and shearing forces of wind.[7] This protection requires compounds such as **lignin** and **cellulose** to stiffen needles against wind and ice and resins to inhibit water loss. It is far more effective to stiffen a needle than to strengthen a flat blade that could be torn and tattered by blowing snow and wind shear. Lignin, cellulose, and resins are carbon rich, and the carbon shunted to them does not participate in photosynthesis. This not only diverts this carbon from the photosynthetic machinery but also makes the needles heavier per unit of the leaf's surface area across which carbon dioxide is obtained from the atmosphere. Photosynthetic rates in conifers are therefore lower per gram of leaf mass and per square centimeter of leaf area than those in deciduous leaves.

Living longer also increases the chance that insects or other animals will find the leaves and chow down on them. To protect themselves against being eaten, evergreens continually maintain active and large pools of compounds such as tannins and phenolics, whose sole purpose seems to be to make the leaves taste bad or to make the animals who eat them ill. These protective compounds take much carbon and energy to produce and do not participate in photosynthesis. They are therefore an additional cost to the tree that could otherwise go into increased growth.

These relationships between leaf chemistry, photosynthesis, leaf lifespan, and habitat are the primary dimensions of the natural history of leaves, especially in the North Woods. These relationships have been recently formalized in an article by Ian Wright and thirty-two coauthors.[8] Wright and colleagues compiled data from 2,548 species and 175 sites around the world. This huge dataset of the natural history traits of leaves suggests that leaf lifespans, photosynthetic rates, leaf area, and leaf chemistry evolved together to produce an integrated suite of strategies ranging from quick (deciduous) to slow (evergreen) return on nutrients and carbohydrate investments. Wright and colleagues call this suite of correlated strategies the Leaf Economics Spectrum.

But not so fast, say Jennifer Funk and William Cornwell.[9] Does it make sense to compare leaf lifespan and other leaf traits between species scattered across the globe? After all, spruce in the boreal forest does not compete with bananas in tropical forests. Perhaps the better scale to examine correlations between leaf lifespan and leaf traits is between species within a plant community, because these plants are competing with each other for nutrients, water, and light. When Funk and Cornwell analyzed the same data that Wright and colleagues collected and looked more closely at these traits within plant communities, the correlations between leaf lifespan and other leaf traits were not always so clear. In communities where there is little variation in leaf lifespan, such as prairies and deciduous shrub communities, photosynthetic rates, leaf chemistry, and leaf area were only weakly correlated, if at all. But in communities with a wide mix of evergreen and deciduous species and hence leaf lifespans, such as the North Woods, photosynthetic rates, leaf chemistry, and leaf area were more strongly correlated. So the correlation between traits that control carbon uptake may depend on what controls leaf lifespan. Funk and Cornwell think that climate may control the variation in leaf lifespans within a community. Climates with small variation between seasons may select for a limited variation in leaf lifespans and hence a narrow range of other leaf traits and weak correlations between them. But climates with more contrasting seasons, such as that of the North Woods, seem to allow species with a wide variety of leaf lifespans and other leaf traits to coexist. In these communities, there are stronger correlations between these traits. So perhaps the way climate controls the assemblage of species and the variation in leaf lifespans drives the correlation between other leaf traits, including rates of photosynthesis and leaf chemistry.

The final word has obviously not yet been spoken about the relationship of leaf lifespans, other leaf traits, and community assemblages. But the pattern seems to depend strongly on the scale over which the data is analyzed, whether global (Wright et al.) or local (Funk and Cornwell). Ecologists need to think carefully about which scales are appropriate for their data and hypotheses; otherwise, conclusions derived from data analyzed at one scale may be inappropriately applied at another.

Things are even more complicated than this. The Leaf Economics Spectrum theory implicitly assumes that leaves act independently of one another. However, leaves do not exist independently of one another but are arranged in a tree's crown, where they modify their light environment and must contend with the changes they have wrought. The Leaf Economics Spectrum must be expanded into a Whole Tree Crown Economics Spectrum to include the correlations between leaf lifespan and crown geometry.

In northern ecosystems, such correlations between leaf lifespan, leaf chemistry, and photosynthetic requirements and rates are especially well developed. These correlations may represent tradeoffs between leaf longevity, photosynthetic rate, and the high costs of constructing a leaf. In a further development of his model, Kikuzawa and his colleague Martin Lechowicz found that the lifetime carbon gain of deciduous and evergreen leaves was approximately the same: High photosynthetic rates and short lifespans in deciduous trees do no better but no worse than low photosynthetic rates and long lifespans in evergreen conifer needles,[10] so no one strategy has a competitive edge in this landscape. The different rhythms of leaf production and lifespan and the diversity of glacial landforms that provide a variety of environments may be what allow evergreen conifers and deciduous hardwoods to coexist in the North Woods. These rhythms are like the point and counterpoint in a Bach fugue: It is difficult to say why it works, but somehow in the end the whole is more pleasing than the sum of its parts.

6.

The Shapes of Leaves

The diversity of leaf shapes and the diversity of ways North Woods tree species capture the sunlight.

Many, perhaps most, of us received our first formal lesson in natural history by making a leaf collection in grade school. I did, in sixth grade. Mrs. Montelewski, our teacher, was captivated by the study of nature. Her science lessons almost always got us out of the classroom and into the forest behind our school. In autumn, we collected and identified the leaves that gravity delivered to us, brightly colored and crisp. The leaves in this little patch of the North Woods behind our school presented an astonishing variety of shapes that we matched to the drawings and descriptions in various field guides from Mrs. Montelewski's personal collection, one of which was Richard Preston's classic *North American Trees*,[1] which is still worth having.

As we saw in the previous essay, the big split in leaf form is between the needle-leaved conifers and the broad-leaved hardwoods. In the North Woods, the needle-leaved conifers are evergreen except for the deciduous tamarack; the broad-leaved hardwoods are deciduous except for a few evergreen bog species such as leatherleaf, Labrador tea, and kalmia. Worldwide, this correlation between deciduousness and broad leaves on one hand and evergreenness and needle-shaped leaves on the other does not always occur; there are many evergreen broad-leaved species in Mediterranean climates, in the Himalaya (e.g., rhododendrons), and in the tropics. But in the North Woods, deciduousness is paired with broad leaves and evergreen is paired with needles, with only a few exceptions.

A hardwood leaf has two parts: the blade, which is the flat surface that captures light and where photosynthesis takes place, and the petiole, which is the stemlet that attaches the blade to the twig. The shape of a hardwood's broad leaf is determined by the nature of the edge of the blade, otherwise known as the leaf margin. Margins of leaves can be smooth or toothed, as well as lobed or unlobed. A toothed margin has small teeth like a saw blade, whereas a smooth margin lacks teeth. A lobed margin has large indentations that extend toward the middle vein of the leaf and divide it into several lobes.

We need to measure how toothed or lobed a leaf is in order to do experiments to test hypotheses about their ecological and evolutionary advantages. "Toothiness" is measured by counting the number of teeth along the entire margin and also by counting the number of teeth per centimeter of margin length. Henry Horn has proposed an elegant way to measure the how deeply lobed a leaf is,[2] and I know of none better. He recommends inscribing the largest circle possible within the leaf blade so that it touches the bottom of the indentations between lobes. Then construct another circle that has the same area as the leaf (I'll explain later an easy way to measure leaf area without an expensive leaf-area meter). The ratios of the areas (or alternatively, the radii) of the two circles is a good measure of the depth of indentations and therefore how lobed the leaf is.

In the North Woods, tree species with toothed leaves include red maple, yellow and paper birch, beech, alder, the various juneberry species, raspberries, blueberries, white and black ash, basswood, quaking aspen, and (of course) big-toothed aspen. Of all the major deciduous

tree species in the North Woods, only sugar maple and red oak have smooth, untoothed margins. All species of maples and oaks have lobed margins, with (red maple) or without (sugar maple) teeth.

Lobes and toothed margins of broad leaves emerged simultaneously with deciduousness in the fossil record about 70 to 80 million years ago, during the late Cretaceous period,[3] in the northern supercontinent of Laurasia, which included North America, Europe, and northern Asia. The rise of deciduousness introduced a new growth strategy and a new manner of competition for light, completely restructured food webs (herbivores had to eat other things during the leaf-free season), and accelerated the cycling of nutrients.

At the same time as deciduousness and toothed and lobe margins were emerging, Laurasia and Gondwanaland, the other supercontinent, which included South America, Africa, India, and Antarctica, were breaking up into the continents we know today. This rearrangement of continents ushered in a prolonged period of global cooling of about 7–10°C, which culminated in the great Quaternary glaciations. The movement of North and South America westward and the collision of India with Eurasia raised the Rockies, the Andes, and the Himalayas along their leading edges. Concurrently, Antarctica drifted to its present location over the South Pole. A large amount of the earth's landmass in these mountain ranges and on Antarctica became covered with permanent snowpacks, glaciers, and ice sheets whose bright white surfaces reflected almost all of the sunlight falling on them, along with its warmth, back into space. In addition, as North America joined South America at the Isthmus of Panama, the flow of warm water between the Pacific and the Atlantic was blocked, causing the formation of the cold Humboldt Current off the coast of South America and creating the Gulf Stream, which brought moisture to northern latitudes, where it fed the growing snowpacks.

The concurrent rise of the great diversity of deciduous leaf shapes and the global rearrangement of continents after the Cretaceous were two of the greatest revolutions in Earth's 4.5-billion-year history. The first changed how energy flows through food webs and the second changed the energy balance of the earth and therefore the climate. The world has just not been the same since the Cretaceous.

Just as toothed margins and lobed leaves arose during a general cooling of the globe, today they are also more common in cooler climates. The relationship is sufficiently strong for paleoecologists and paleoclimatologists to calibrate the percentage of toothed leaves in a plant community to mean annual temperature and then use this to back-calculate paleotemperatures from fossil leaf assemblages.[4] Where mean annual temperatures are between 4°C and 10°C, which is the range of temperatures of the North Woods, fully 80 percent or more of deciduous species have leaves that are toothed, more than in all other flora of warmer climates. In the leaves of North Woods species, there are also more teeth per centimeter of perimeter (about three) and more teeth per leaf (about fifty), and tooth area is a greater proportion of blade area (about 5 percent) compared with the leaves of warmer climates.

The evolutionary appearance of toothed margins and lobed leaves during the late Cretaceous cooling and the predominance of toothed and lobed leaves in regions of cooler mean annual temperatures today suggest that there may be some evolutionary advantage to teeth and lobes in cool climates. One hypothesis is that toothed and lobed margins have high photosynthetic rates early in the spring, resulting in a longer growing season than that of leaves with smooth margins and without lobes. In an elegant experiment testing this hypothesis, Kathleen Baker-Brosh and Robert Peet exposed newly emerged leaves collected in spring from a wide variety of eastern deciduous species with toothed, lobed, and smooth margins to radioactive $^{14}CO_2$ for 30 seconds.[5] This was a short enough time to take up the $^{14}CO_2$ and incorporate it into sugars but not long enough for the sugars to be translocated throughout the leaf blade. After the leaves were harvested they were immediately dipped in liquid nitrogen to stop any additional translocation of sugars. Baker-Brosh and Peet then laid the leaves on photographic film and allowed the radioactive ^{14}C to blacken the image of the leaf in the spots where photosynthesis made it into sugars. In more than half of the toothed or lobed species, these spots were in the teeth and apices of the lobes. Leaves with smooth margins lacked spots of concentrated ^{14}C. Clearly, photosynthetic activity during the early spring emergence of leaves is much greater along toothed and lobed margins than in leaves with smooth margins or without lobes. Baker-Brosh and Peet call these toothed and lobed margins "precociously photosynthetic."

But, there's a catch. These higher rates of photosynthesis in toothed and lobed margins come at a cost of greater water loss by transpiration compared with leaves with untoothed margins or without lobes.[6] The early season loss of water may not be that great a cost in northern climates compared with the warmer South because by early spring snowmelt has usually recharged northern soils. Significantly, southern forests have far smaller percentages of species with toothed and lobed margins, and toothed and lobed leaves are also less abundant in water-stressed environments.[7] Water losses through the teeth are the price to be paid for rapid photosynthesis in early spring; this price partly dictates the habitats where these species can thrive, namely the fine-textured soils that can hold water found mainly in tills such as end and ground moraines.

Even within the same species, populations in colder environments are toothier and more deeply lobed than populations in warmer climes. How much of this is due to genetic differences between populations and how much is simply adjustment of the tree to the local environment? To answer this, Dana Royer and colleagues grew red maple seedlings collected from North Woods populations and from southern populations in gardens in Rhode Island and Florida.[8] This is known as a common garden experiment because individuals from geographically distant populations are planted in common in gardens located throughout the range of the experimental material. A common garden experiment allows one to separate the genetic component of a trait (by comparing traits from different populations in the same garden) from the adjustment of the trait to the local environment of the garden (by comparing the trait from each population in different gardens). The northern populations of red maple had leaves with more teeth and more deeply dissected lobes than the southern populations, regardless of which garden they were grown in. This is the genetic component of leaf shape. But plants from northern populations also produced fewer teeth and smaller lobes when planted in the southern gardens than when planted in the northern gardens. This is the adjustment of the leaf shape of trees from northern populations to the local climate in which they are grown. Royer and colleagues calculate that 69 to 87 percent of the geographic variation of leaf shape in red maple was caused by genetic differences between local populations, and 6 to 19 percent was due to adjustment of a population to local environments. Whether the predominance of genetic control over local adjustment holds for other species is not yet known.

The increase of toothed margins and deep lobes with cooler temperatures at higher latitudes does not hold, however, for the distribution of
toothed and lobed leaves within a tree's crown. Although temperatures
decrease into the canopy, the shade leaves there are less lobed and have
smaller teeth than the sun leaves in the warmer temperatures higher
in the canopy, the opposite of the correlation with latitude. This is a
striking asymmetry in the response of leaf shape to temperature. Why
should teeth and lobes become more prevalent in cooler climates in
northern regions but less prevalent in the shaded and cooler interior of a
canopy? After all, the leaf doesn't "know" whether the cool air surrounding it is because of the latitude or because of shade. Maybe temperature
is not, after all, a strong or even the only controller of leaf shape. Could
leaf shapes be responding to something other than gradients of temperature across latitudes and within the canopy? Perhaps leaf shape is
controlled by light. The total amount of light that impinges on a deciduous leaf over its lifetime increases both upward through the canopy
and with longer growing season daylengths in northern latitudes. The
propensity toward more teeth and deeper lobes in more sunlit regions
both higher in the canopy and where growing season days are longer
suggests that perhaps leaf shape is responding to light gradients rather
than temperature gradients. But this contradicts the long-term evolutionary development of teeth and lobes during global cooling after the
late Cretaceous period. There is a paradoxical asymmetry here in how
leaf shape responds to temperature and light at global, continental, and
canopy scales and also over evolutionary times compared with the lifetime of an individual leaf. This paradox begs to be resolved. Paradoxes
uncovered by natural history observations such as these provide some of
the strongest impetuses for further research.

How does a leaf "know" where to make a tooth or lobe, and how
many teeth and how large a lobe to make? There must be physiological
and molecular regulatory mechanisms responsible for controlling leaf

shape. At the molecular level, sites of tooth and lobe formation are also sites of high activity of the growth hormone **auxin**.[9] It is therefore a reasonable hypothesis that auxin controls the development of teeth in leaves. If this hypothesis is correct, then shutting down or slowing auxin production in toothed margins should attenuate the growth of teeth. As far as I know, this experiment is yet to be done.

What would a leaf look like that was all tooth and no interior blade? It would look like a conifer twig with its needles radiating out from it like long, skinny teeth. We can think of the current year's production of needles on a conifer twig as one leaf that is all teeth. How does the surface area of the needles on a current year's twig of a conifer compare with that of a broad deciduous leaf growing in the same environment? To find out, I sampled the needles on four twigs from a white spruce containing only this year's needles and four leaves from a nearby sugar maple at the end of the growing season when all the needles and leaves were fully expanded. Both trees were growing on the same soil, so I could control for soil fertility, and both were large, mature trees, so I could control for size and presumably age (more or less). I collected the leaves and twigs from the same height and orientation in the crown, so they were exposed to approximately the same light levels. I measured the width and length of all spruce needles from each twig with a digital micrometer. Assuming the flat side of a needle is approximately a long skinny rectangle and knowing that spruce needles have four sides, I could calculate the average surface area of the needles and then multiply it by the number of needles on each twig to get the total surface area of this year's needles produced on each twig. To calculate the surface area of the sugar maple leaves, I could have used a leaf-area meter but instead did it the old-fashioned way by copying each leaf on a sheet of paper with a photocopier, cutting the shape out, and weighing it. From the weight and area of a sheet of paper I could calculate the area of each leaf by simple ratios.

The average width of these spruce needles was 1 millimeter, the average length was 19.6 millimeters, the average area of the needles was 78.3 square millimeters, there were an average of 96 needles per twig, and the average needle surface area produced in 1 year's growth of a twig was 7,536 square millimeters. This was almost identical to the average area of 7,854 square millimeters of each of the four sugar maple leaves. Now, this is a small sample, and a more rigorous test of the hypothesis requires more samples from different positions within the canopies and from trees growing on different soils. Nevertheless, the close correspondence between surface areas of the spruce needles and maple leaves growing under reasonably similar conditions is striking. The photosynthetic unit of a conifer that is analogous to a deciduous leaf may not be a needle but all the needles on the current year's growth of a conifer's twig.

Even though the leaf area produced by this year's growth of a spruce twig and a maple leaf are similar, the entire conifer tree has at least twice as much leaf area as a deciduous tree because northern conifer needles are retained for several years. This means that conifers have more layers in their crowns than deciduous trees, but this poses the problem of how to get light into the older leaves inside and lower in the crown. This is where the small size of the conifer needle is at an advantage. Small needles cast smaller shadows than large broad leaves, so more small needles can be packed closer together into the volume of the canopy without shading each other too much. The densely packed needles along a conifer shoot can also be effective adaptations to cold temperatures because they slow the flow of air and thereby decrease convective heat losses. · This can raise temperatures around a conifer twig well above freezing even during winter, allowing photosynthesis to proceed before spring leaf flush of deciduous trees in the same stand.[10] Needle-shaped leaves therefore allow the evergreen strategy to be successful in northern environments by extending the photosynthetic season and hence the growing season, almost year-round.

The stiff needle shape also gives conifers greater control over leaf angle than the floppy leaves of deciduous broad-leaved species. Doug Sprugel claims that this allows conifers to angle their needles to make the most effective use of the light dispersed throughout the crown.[11] Needles on the outer rims of conifers maximize their photosynthetic rates at light intensities of 25 percent of full sunlight. This is approximately the amount of light a needle would receive if it were angled at a shallow 15 degrees to the angle of incoming sunlight. Most needles on the outer surface of a conifer crown are indeed at these shallow angles to incoming sunlight. Capturing more light at angles more perpendicular to sunlight is wasteful for a needle on the crown surface because it would not increase photosynthetic rates. But this leaves 75 percent of the incoming light to disperse through the rest of the crown. This attenuated light inside the crown or near the forest floor is more effectively captured by other needles oriented at angles more perpendicular to the sun's rays. By capturing more of the light that reaches them, these interior needles can maximize photosynthesis even though they are at lower light levels. In Sprugel's phrase, orienting needles in the outermost and uppermost layers of the crown at shallow angles to incoming sunlight "spreads the wealth" among all needles.

Although much has been learned about the evolution and adaptive significance of leaf shapes in both conifers and deciduous species, much remains to be explored. Comparative studies between well-chosen species differing in only a few characteristics of leaf shape are greatly needed. For example, a comparative study of photosynthesis and transpiration throughout the canopies of red maple, which is lobed and toothed, and sugar maple, which is lobed but not toothed, along various environmental gradients would be interesting. Possible connections between leaf shape and carbon and nitrogen economies also have not been explored in detail, even though leaf shape may be a part of the Leaf Economics Spectrum (see Essay 5). For example, rapid photosynthesis and growth of teeth in angiosperms require large amounts of nutrients, especially nitrogen, and species with large amounts of tooth area per centimeter of perimeter also have high nitrogen contents.[12]

Perhaps some child in grade school who is now making a leaf collection will someday answer these and other questions and thereby deepen our understanding of the evolution, physiology, and ecology of leaf shape.

7.
The Shapes of Crowns

The North Woods has a very high diversity of crown shapes, from conical conifers to globular hardwoods. What are the advantages and disadvantages of these crown shapes, and how do they help the different species survive in different places in the landscape?

As I look out my window at the skyline of the ridge a few hundred meters away, I see a profusion of crown shapes. These include the rounded and vase-shaped crowns of maples, the wide and spreading red oak, the elliptical crowns of birch and aspen, the club-shaped but more conical white spruce, the Christmas tree–shaped balsam fir, and the white pines shaped like candelabra and towering above the rest, all of them reaching for the sun.

The sunlight dances down through each crown in a different way. Some of the light captured by a leaf propels electrons through the leaf's photosynthetic machinery; the rest is deflected to other leaves. Each crown controls the dance of sunlight by the shape of its outermost surface and by how leaves are distributed in its interior. Eventually, the energy of the sun reaches the forest floor either as a sunfleck or, when captured by photosynthesis, in chemical bonds in the fallen leaves.

To the south of the North Woods, in the Eastern Hardwood Forest, most tree crowns are broad and rounded. As one passes through the North Woods to its cousin the Boreal Forest to the north, the forest canopy becomes increasingly dominated by the conical crowns of spruce and fir. This segregation of crown shapes across the surface of the earth remains one of the biggest mysteries in plant biogeography since the beginnings of modern ecology.[1] How can we make sense of this segregation of crown shapes at different latitudes of the earth's surface?

Like leaf shapes, crown shapes are evolutionary experiments in the geometry of how light interacts with photosynthetic surfaces. In making a crown, a tree has to solve the problem of arranging leaves to capture as much light as possible without shading each other, while simultaneously minimizing water loss. The crown shapes that survive in different climates or on different soils are the successful evolutionary solutions to this problem.

Can we explain the biogeography of crown shapes by the physiological and geometric properties of the crown and leaves? To address this question, Yossi Cohen and I constructed a mathematical model of tree growth that included the shape of the crown, photosynthetic rate, and the average density of leaves in a crown, as well as other properties.[2] The trick to constructing this model was to imagine each crown as a set of infinitely thin disks of different radii stacked atop one another. The shape of this stack of disks is controlled by the ratio between the radius and the depth of each disk. As we go deeper in the crown from its top, this ratio determines what the radius of the disk at that depth will be. A large ratio makes the radii of the disks grow rapidly with depth and therefore produces a wide, umbrella-like crown, as many deciduous hardwoods have. Intermediate values of the shape ratio produce conical crowns such as those of pines and balsam fir because the radii increase by a certain and constant percentage with every increment of depth. Small values produce disks whose radii barely grow with depth and therefore result in candle-shaped crowns such as in black spruce trees in the Far

North. We concentrated on how the shape of the outermost surface controlled tree fitness by assuming that leaves were evenly distributed in the interior of the crown.

Species in the model can evolve by adjusting this ratio or any of the other natural history parameters in order to maximize photosynthesis. Adjusting the ratio affects the degree of exposure to sunlight, which is maximized when the outer surface of the crown is perpendicular to the angle of sunlight. The model predicts that in order to maximize exposure to the sun along a line of longitude, the plant crown should change from the shape of a nearly flat umbrella in the south, where the sun angle is closer to vertical, to a narrower and narrower cone and eventually a candle in the north, where the sun angle becomes increasingly horizontal. Because the outer surfaces of conical crowns are perpendicular to the angled rays of the northern sun, conical tree canopies capture more light at northern latitudes and consequently should have greater photosynthesis and seed production. Trees with the genes that produce conical crowns will therefore leave more descendants in the north than trees with genes that produce flat, umbrella-like crowns. This is the essence of Darwin's theory of natural selection. So the distribution of different crown shapes northward is controlled by the angle the outermost surface makes with the rays of the sun.

The tree in the model is also free to vary any of the other parameters in addition to or instead of the ratio. For example, the tree can also decrease the density of leaves in the crown to minimize self-shading and thereby also increase photosynthesis within lower disks without changing crown shape. The evolutionarily optimal solution for trees is not a single value for any one parameter but instead a coordinated and correlated set of parameter values that together maximize photosynthesis. But the tree can vary the photosynthetic rate and other physiological parameters only within narrower biochemical limits. In contrast, crown shape can be changed over a wider range of values and thereby have larger effects on photosynthesis than can the restricted changes in the biochemical machinery of photosynthesis.

What about how leaves are distributed throughout the crown? We assumed that leaves are evenly spread throughout a crown, but stand beneath any tree, look up, and you will see that this is clearly not always the case. Does it make a difference? Are there optimal distributions of leaves through the crown for different environments?

These questions were elegantly addressed by Henry Horn in his classic *The Adaptive Geometry of Trees.*[3] Horn proposed that a tree can arrange its leaves in one of two ways. At one extreme, a tree can put all its leaves in a single layer in the outermost surface of the crown and have each leaf touch adjacent ones, thereby capturing all the light imping-ing at full strength on its crown. Horn called this leaf distribution a monolayer. But by placing all its leaves in full sunlight the tree is also burdening them with a high heat load and therefore a high loss of water through transpiration. Monolayered crowns pay the price of placing all their leaves in a single brightly lit layer by having their habitat restricted to sites of adequate to high moisture with low probabilities of drought. In the North Woods, these sites are predominantly on the clay-rich soils of moraines and the sides or bottoms of kettles.

Alternatively, the tree could distribute all its leaves randomly through-out its crown, but then some leaves would be at least partly shaded by those above them. Horn called this a multilayer. The tradeoff is that although the shading reduces photosynthesis to some degree, it also reduces water loss, and so multilayers should be able to survive drought. This gives them a competitive edge over monolayers on the drier sandy outwash plains the ice sheet left behind. Still, the tree pays a price by having some leaves at least partly shaded because some overlap of leaves is inevitable with a random distribution. This is less of a problem than it may seem if the tree exploits two loopholes. The first loophole is that photosynthesis in most species approaches a maximum rate (the term *saturates* is often used), which happens between 20 and 30 percent of full sunlight. Any more sunlight than that is wasted. The second loophole is that the shadow of a leaf does not extend completely to the ground but is instead an umbra, or a wedge of partial darkness on the opposite side of the leaf from the sun. The length of the umbra is between fifty and seventy leaf diameters behind the leaf surface. So leaves within the crown can still maintain nearly maximum photosynthesis if the upper layers maintain a distance between them of fifty to seventy leaf diameters, collectively allowing at least 20 percent of full sunlight to pass through.

Monolayers and multilayers are two extreme endpoints of crown structure, and most trees lie somewhere in between. How can we measure how many layers the crown of a tree or species has? Horn mathematically derived a formula that calculates the number of layers in a crown from the ratio of the logarithm of the proportion of light penetrating the entire crown to the logarithm of the proportion of light penetrating a single layer. He assumed, reasonably enough, that a single horizontal branch constitutes a single layer. These amounts of light can be measured by holding a light meter beneath the crown for the entire tree and holding a light meter above and below a branch. This latter requires an arborist's or orchardist's ladder or climbing equipment to measure light capture through a branch. But there is an easier way to get good approximations for these measurements, and that is to sight upward through a cardboard tube aimed through the crown or through a well-chosen branch not obscured by other branches, then estimate the proportion of the visual field covered by sky. This estimate is proportional to the amount of light penetrating the crown or branch. Horn says that with practice, one can calibrate one's eye to make this estimate fairly accurately. Horn tested this method with students and found that although their raw data of estimates of proportions were "appallingly disparate," the ratios of the logarithms required by the formula were surprisingly consistent.

Using this method, Horn estimated that on average birch, aspen, and white pine had 2.5 layers, red maple and red oak had 2.1 layers, and sugar maple and beech had 1.5 layers. This sequence of species should be familiar as the classic successional sequence in the North Woods, from aspen or birch dominating a forest after a fire or logging, followed by oak, beech, and sugar maple as they grow up from seedlings in the understory and replace the mature aspen and birch when these species die. So during succession from aspen and birch to maple, oak, and beech, the number of layers in the crowns of the dominant tree species decreases.

What are the advantages of multilayered crowns shortly after a disturbance and monolayered crowns late in succession? In the harsh and often dry environment of a clearcut or fire, the early successional multilayered birch, aspen, and pine have an advantage over the monolayered and drought-intolerant sugar maple and beech. The multilayered crown allows these species to survive water stress because of the reduced heat loading on the partially shaded leaves in the lower layers of the crown. But the mutual shading of leaves in a multilayer probably doesn't matter so much in the full sunlight of an open clearing because all the leaves in the crown receive enough sunlight to maintain near maximum photosynthetic rates. Surviving water stress is a good thing for these species because their light seeds are widely dispersed and must make do with whatever soil moisture they find wherever they land, which could include very dry soils. But the price these shade-intolerant species pay is that they cannot reproduce in their own deep shade because, by the mutual shading of its own leaves, a multilayered sapling or seedling in the understory exacerbates the already reduced sunlight penetrating the overstory.

In contrast, the monolayered crowns of shade-tolerant sugar maple and beech enable their seedlings to capture what little light penetrates the upper layers of the forest canopy without shading leaves within their crown. Populations of sugar maple and beech can thereby maintain themselves by having seedlings survive in the understory of the so-called climax forest. These species do not have to disperse their seeds to sites with full sunlight and therefore can produce heavy seeds that contain sufficient food reserves to get their seedlings started. The price they pay is their high need for a steady supply of soil moisture, because when the seedlings reach the overstory, all their leaves are on the surface of their monolayered crowns and are all suddenly exposed to the high heat loading of the full sunlight.

The seedlings in an understory of a forest are not bathed in attenuated sunlight but instead in a mosaic of sunflecks and darkness. The majority of understory forbs and seedlings of canopy trees achieve most of their photosynthesis only in the brief moments when sunflecks pass over them.[4] What is important to the understory seedlings is not so much the average amount of sunlight on the forest floor but its statistical distribution in sunflecks.[5] Although these statistical distributions of sunflecks are easily measured by a variety of methods,[6] much remains to be learned about them. As I walk through the various stands in the teaching forest on my campus, I can see that the size and spatial distributions of sunflecks differ markedly depending on the shape and distribution of leaves within the crowns. How does the survival of seedlings relate to the distribution of dark times between sunflecks as the sun passes overhead, during which photosynthesis is very low? Answers to these questions might help us understand more about how survival of seedlings relates to the adaptive geometry of the crowns of the adults that are their parents.

Does Horn's theory describe the distribution of canopy structures of entire forests in addition to the crowns of individual trees along a moisture gradient? John Aber and I set out to test this hypothesis by measuring the vertical distribution of leaves in forest canopies along a soil moisture gradient in Wisconsin from very dry sands on outwash plains to silty clay loams with abundant soil moisture reserves on moraines.[7]

To measure the vertical profile of the canopy we used an old-fashioned Pentax single lens reflex camera with a 120-millimeter focal length telephoto lens mounted vertically on a tripod, a camera that is becoming rare with the advent of digital cameras and even cell phone cameras. We replaced the focusing screen in the viewfinder with a screen that had fifteen grid points that we could use to focus the telephoto lens on fifteen individual leaves. We could then read off the distance to the leaf from the camera's rangefinder, which we had more precisely calibrated using a tape measure stretched out on the ground. By getting measurements to the heights of fifteen leaves at many locations within a stand, we could accurately characterize the vertical profile of the canopy.

We found that Horn's predictions of change in crown structure from multilayers to monolayers with increasing soil moisture held for the entire forest canopy only from the middle of the soil moisture gradient to the most moisture-rich end. On the moraines with abundant moisture held by the silty clay loam soils, the forest canopy was a tall, high, almost single layer of leaves dominated by sugar maple. The canopy of the forest as a whole was a monolayer just like the crowns of individual sugar maple trees. On moraines whose soils were sandy loams that could hold less moisture, the overstory became dominated by red oak, with a multilayered crown overtopping shorter red maple trees lower in the canopy. The distribution of foliage from the top of the canopy to the ground was therefore also a multilayer. On very dry sites, red maple dropped out of the plant community, and the upper canopy was dominated by the open crowns of large, isolated white oaks that transmitted sufficient light to a short but continuous shrub layer beneath them. The driest sites therefore had forests with double-layered canopies, a type not predicted by Horn's theory for crowns of individual trees. Successive changes of species with different crowns along soil moisture gradients therefore produce changes in whole forest canopies, from monolayered to multilayered to the midpoint of the soil moisture gradient, but then produce a double-layered canopy of sparse trees over pronounced shrub layers on very dry sites, which does not correspond to the crown of any individual tree.

The increasing predominance of conical crowns through the North Woods and the diversity of leaf distributions within the crowns result from selection for shapes that maximize exposure to the angles of sunlight while minimizing water loss. The diversity of crown shapes in the North Woods, each of them apparently as successful as the rest, may result from the coincidence of the current distributions of its constituent species along the 45th to 48th parallels of longitude and from the high diversity of soils of different water-holding capacities on moraines and outwash plains left by the ice sheet. At these latitudes, the angle of sunlight begins to change from having a strong vertical component, which favors rounded crowns, to a strong horizontal component, which favors conical crowns. The diversity of soils provides a landscape where both monolayered and multilayered crowns can find adequate habitat. Perhaps in these latitudes and across the diverse glaciated landscape of the North Woods, no one crown shape greatly outperforms the others in capturing the light.

There is exquisite poetry in the visual diversity of the crowns and canopies of the North Woods and in the mathematical way that different species adapt their crowns to capture light. I think this would have been appreciated by Jun Fujita, a Japanese poet who retreated to the North Woods of Minnesota, where he built himself a cabin and lived between 1928 and 1941.[8] Fujita was the first poet to adapt the ancient Japanese *tanka* poetic form to English (*haiku* is the first three lines of *tanka*). Perhaps the canopies of maple and spruce inspired his poem *Morning Woods*, which contains the following lines:

A static mood, in the morning woods,
Wet and clear—
In a majestic pattern, leaves are spellbound.[9]

8.
How Should Leaves Die?

Exploring the processes by which a leaf dies and how these determine its decay and the cycling of nutrients.

It is only the beginning of August, but as I pick blueberries, I already see a red leaf here and there on the bushes. In the next 60 days, the crescendo of leaf color will herald the shift from the brief summer of the North Woods to its long winter.

As each species has its own characteristic leaf and crown shapes, so does it have its own characteristic color. The reddest leaves are produced by blueberries and bearberries, which carpet the understory with crimson. Of the trees, the maples turn first, and their hues can range from bright red to brilliant orange. Birches and aspen begin to turn yellow about the time the oranges and reds of maple peak, but the yellow of birches is usually opaque, whereas the yellow of aspen glows transparently from within. Of the broad-leaved species, red oaks turn color last. Theirs is a dusky red, earthier than the flame red of maples. The needles of larch, one of the few deciduous conifers, flame bright yellow in the peatlands in early November. Older needles of the pines, spruces, and firs also turn yellow before dropping in autumn, but these species retain their summer's load of fresh, deep green chlorophyll in their younger needles. A green balsam fir against a red sugar maple is a harbinger of the coming celebrations around the winter solstice.

The characteristic colors of the dying leaves of different species emerge from their individual chemistries. In summer, an abundance of chlorophyll floods the leaves of all species. The green we see is the color of chlorophyll, which absorbs wavelengths of light in most colors but reflects the wavelengths of the green portion of the spectrum. All other wavelengths of light are absorbed by chlorophyll and thus drive photosynthesis, including the longer, near-infrared wavelengths that lie just outside our eye's ability to detect.

Chlorophyll is not the only pigment in cells or even the only pigment in the photosynthetic machinery. The other major pigments are in the families of carotenes, xanthophylls, and anthocyanins, which assist chlorophyll in the photosynthetic machinery. Carotenes are orange pigments that absorb additional light and transfer it to chlorophyll. Excessive light can excite chlorophyll to a very high energetic state where it cannot participate in the photosynthetic machinery, and the highly excited chlorophyll can then damage the cell. The yellow xanthophylls help dissipate this excess energy as heat, which warms the leaves' surroundings. Anthocyanins are red, purple, or even blue pigments that absorb the very energetic wavelengths of ultraviolet light and protect chlorophyll and leaf cells from damage. Anthocyanins are the leaves' sunscreen.

Orange carotenes, yellow xanthophylls, and red anthocyanins are the pigments that we see in autumn when the waning daylight signals the leaves to slow the production of chlorophyll. These other pigments are then unmasked in the absence of chlorophyll. In fact, our eyes are always detecting the orange, yellow, and red wavelengths reflected from these pigments, and so in effect we are "seeing" them, but the amount of green wavelengths reflected by chlorophyll in summer overwhelms our perception of the canopy's other colors.

Some years the colors are intense, and we call it "the best year ever," but in other years the colors are washed out and dull. Our inability to exactly predict how intense the colors will be, especially how bright the reds and how deep the oranges will be, is what keeps us anticipating the autumn color changes anew each year. "How do you think the colors will be this year?" "Are the colors good in your part of the woods this year?" "Too bad about the fall colors this year." Every year, state departments of conservation or natural resources from Minnesota to Maine post maps on their websites delineating the timing and intensity of fall colors; these maps are updated almost daily and are some of the most viewed portions of these agencies' websites.

The intensity of fall colors, especially the reds, depends on how good the summer growing season was. Sugars are precursors to the red anthocyanins, and their production is increased by dependable soil moisture, especially during the growing season. Leaves are especially high in red anthocyanins at the end of a good growing season that had plenty of available water in the soil to increase sugar production. These are the years when we get the deepest red colors that everyone likes the best. In contrast, after summer droughts fall colors are usually drab because of the lack of sugar production.

If bright colors require regular supplies of water in the summer, they are also enhanced by clear blue skies, dry weather, and cool temperatures in autumn. As the leaves turn color, the integrity of their cell walls breaks down. Rainwater washing over leaves in autumn can wash these pigments out, leaving a drab skeleton behind. Put a clear glass jar beneath a sugar maple in autumn and collect some rainwater washing off the canopy. The rainwater will take on the colors of the overhead leaves. It will also have a slightly sweet taste from sugars washing out along with the pigments.

The death of leaves in autumn does not happen simply because a frost kills the leaf tissue, as is commonly thought. A better term than leaf death is leaf senescence. Leaf senescence is a genetically programmed set of processes that form several layers of special cells at the base of the petioles of the leaves or needles where they are attached to the twig.[1] Whereas senescence is the overall decline and death of the leaf, including the changes in color, abscission is more specifically the separation of the leaf from the twig. These layers of cells at the base of the petioles consist of a layer toward the leaf blade where the leaf separates when it falls and a second corky and protective layer toward the twig. The corky protective layer stays on the twig and forms a scar with a shape characteristic of each species, so characteristic that the species can be identified in winter even without their leaves just from the scars on the bare twigs.[2] The separation and corky layers form because environmental cues such as falling temperatures and reduced daylength trigger some as yet unknown complex of genes to alter the growth of their cells. Which genes are responsible for initiating the formation of separation and corky layers and how the genetic system translates environmental cues to their development is almost entirely unknown at present. As abscission proceeds, the cells in the separation layer become less strongly attached to each other, and eventually the leaf detaches along a single layer of cells and falls to the ground.

In some species, most notably red oak, the process of abscission begins but then stops midway through the formation of the corky and separation layers. This arrested abscission is known as marcescence. Marcescent leaves remain green later in the fall than leaves undergoing normal abscission, so marcescence may be a way to extend the growing season by a few precious weeks. Leaves on trees with marcescent abscission are retained throughout the winter, turning deeper and deeper brown as melting snows wash out the red anthocyanins and the other pigments. In spring, abscission resumes; and corky and separation layers are formed, and leaves from these trees fall in the next spring.

Discarding a leaf is a costly process to the tree, especially because

green leaves are its most nutrient-rich part. To save some of these nutrients, almost all plants recover some of the nutrients from their senescing
leaves before they fall in a process known as resorption. During resorption, nitrogen and phosphorus are pulled back out of the leaves and are
stored in the twigs during winter. The stored nitrogen and phosphorus
are then moved out of the twigs into the flush of new leaves next spring,
before the soil thaws and nutrient uptake can begin. Resorbing nitrogen
and phosphorus from senescing leaves and storing them in twigs allows
the tree to extend the growing season in spring by a few weeks.

The ecological importance of resorption was first realized by Doug
Ryan and Herb Bormann in the North Woods at the Hubbard Brook
Experimental Forest in New Hampshire.[3] By comparing the chemical
contents of green and abscised leaves of sugar maple, beech, and yellow
birch before and after abscission and correcting for the amount of nutrients leached from leaves by the autumn rains while they were still on the
tree, Ryan and Bormann found that resorption accounted for about one
third of annual nitrogen and phosphorus requirements of these trees.

Soon after Ryan and Bormann's article was published, ecologists
began to speculate about the evolutionary and adaptive significance of
resorption. On infertile soils, the nitrogen and phosphorus resorbed
during senescence can make up a substantial portion of annual nitrogen
and phosphorus needs of the tree. It would seem reasonable that such
an important mechanism of conserving nutrients would greatly improve
fitness, especially because nitrogen and phosphorus are usually in short
supply from the soil. A common hypothesis was that the efficiency of
resorption would be greater on infertile sites because that would lessen
the tree's dependence on nutrient supplies from the soil. Resorption
efficiency was defined as

$$\frac{\textit{Nutrient content of green leaf - Nutrient content of abscised leaf}}{\textit{Nutrient content of green leaf.}}$$

Despite the reasonableness of this hypothesis and the precise mathematical definition of how to calculate resorption efficiency, no clear pattern merged across numerous studies. Sometimes resorption efficiency increased with soil fertility, sometimes it decreased, and sometimes there was no relation. Sides were drawn, and the argument became spirited.

Part of the problem in this debate is that resorption efficiency, like all measures of efficiency, is a ratio. Ratios are tricky things; you have to watch them like a hawk. A ratio can change because the numerator changes or the denominator changes, or both change at different rates or in different directions. The terms in the numerator and denominator of resorption efficiency are affected by distinctly different biological processes, which are often under the control of different genes. There is no single "efficiency" gene that controls this ratio. Instead, there are processes that control the nutrient content of a green leaf in both the numerator and denominator of this equation, such as soil fertility and uptake rates. There are other processes that control the nutrient content of the abscised leaf, which is only in the numerator, such as the processes that move nitrogen and phosphorus out of the leaf back into the twig. So we have three terms in this equation, each of which can go up or down and shift this ratio one way or another depending on how the other terms behave.

Another part of the problem was the source of the data used to evaluate this hypothesis. Some investigators used primary data that they themselves collected in the field,[4] as Ryan and Bormann did. Others mined data from studies published in journals or in online databases.[5] Many of these studies lacked data on soil fertility. In these cases, the investigators flirted with tautology by using green leaf nutrient concentrations to infer soil nutrient availability. This is a tautology because the green leaf nutrient concentration is already part of the equation used to measure resorption efficiency, so it is improper to use it to also estimate soil fertility.

An additional problem with the literature data is that the sites and species in previously published studies were not necessarily chosen to test the hypothesis of soil fertility controlling resorption efficiency. They were usually chosen to test other hypotheses, and so the sites may not be the best sites to test this particular hypothesis. Moreover, despite the large size of these databases, collectively they are not a random, unbiased sample of the world's ecosystems. Other factors such as the history of the site, drought, insect attacks, and diseases could also affect resorption and therefore confound the results but were often unreported. Literature and online data are very coarse instruments with which to test hypotheses because we might conclude one thing from them when in fact we might conclude the opposite from a carefully controlled and executed field study in which other confounding factors were accounted for. Literature and online databases never (in my view) replace the power of well-designed field experiments. Unfortunately, with research funds being increasingly difficult to procure and with the increased pressure for young scientists to publish to get a job interview or to obtain tenure, using compilations of data from the literature or online databases becomes an attractive alternative to spending years getting a grant and then working several more years in the field to test hypotheses more rigorously.

Keith Killingbeck tried to circumvent some of these problems by proposing that we instead focus on resorption proficiency rather than efficiency.[6] He defined *resorption proficiency* as the lowest concentration to which a nutrient can be reduced in leaf litter by resorption. This definition avoids the problems of the efficiency ratio because resorption proficiency can be measured directly by laboratory analyses of nutrient concentrations in litter tissue, whereas resorption efficiency is a derived piece of calculated data using the equation presented earlier. Changes in resorption proficiency are therefore much less ambiguous than changes in resorption efficiency. Killingbeck augmented his own field data with data from the literature, so his study was not entirely free of some of the problems just mentioned, but he was able to show that evergreen conifers, which usually dominate infertile sites, had lower phosphorus and nitrogen concentrations and therefore greater resorption proficiencies than the deciduous species that dominate fertile sites.

But is greater resorption proficiency an adaptation to low soil fertility, or is it a cause of it? Greater resorption proficiency means decreased nutrient returns to the soil because nutrient concentrations in the leaf litter are reduced. If nutrients are being taken up from the soil and then retained in the tree by resorption before the leaf falls, doesn't that partly deplete the soil of nutrients? Yes, to a limited extent. But there is an even bigger problem, which is that the nutrients bound up with carbon in molecules in leaf litter must be released by microbes and fungi into the soil solution before plants can take them up again. To a great extent, soil fertility depends on the ability of microbes and fungi to break the carbon–nutrient bonds. The microbes and fungi then oxidize carbon to carbon dioxide to obtain energy, much as we do to the carbon in our foods such as the sugar in a candy bar. Simultaneously, the microbes and fungi also release the nutrients from the litter into the soil solution, where they can be taken up again by plants. The forest floor is not merely a pile of dead leaves but a rich living system of microbes and fungi that gain their energy by breaking carbon bonds in leaf litter.

Microbes and fungi have very high nutrient concentrations in their bodies, much higher than the concentrations in leaf litter. Therefore, greater resorption proficiency creates litter that is more deficient in nutrients relative to the demands of microbes. Resorption of nutrients from abscising leaves deprives the microbes of those same nutrients. Decomposition rates are especially slow in conifer needle litter, with greater resorption proficiencies and therefore low nutrient concentrations, compared with deciduous hardwoods, with lower resorption efficiencies and consequently higher nutrient concentrations.[7] The soil humus derived from conifer needle litter also decomposes and releases nutrients more slowly than humus derived from hardwood litter.[8] Soils beneath conifers are therefore less fertile than soils beneath hardwoods.

So greater resorption proficiency and the evergreen growth habitat of conifers together conserve nutrients in live plant tissue, but with the tradeoff of depressing soil fertility through their nutrient-poor and slowly decomposing litter. Deciduous hardwood species such as sugar maple and birch, with lower resorption proficiency, increase soil fertility by boosting microbial activity with their nutrient-rich and rapidly decomposing litter. But this enhancement of soil fertility by deciduous species carries with it the tradeoff of not conserving as many nutrients in their twigs for later use. The chemistry of abscised leaf litter therefore links plant growth and soil nutrient availability in a positive feedback loop. Less proficient resorption means nutrient-rich, rapidly decaying litter, which boosts soil fertility, which then reduces selection pressure for more resorption. More proficient resorption means slowly decomposing litter, which decreases soil fertility, which then increases selection pressure for more resorption. How evolution operates within these feedbacks of the nutrient cycles between plants and soils is currently a wide open question.[9]

As we have seen in previous essays in Part II, the energy of light enters the ecosystem when it is captured by leaves arranged in various geometric patterns in the canopy and with different lifespans and shapes. The energy that comes from sunlight is captured in carbon bonds by photosynthesis and then transferred through the food web when live green leaves are consumed by herbivores or when the dead leaves are consumed by microbes. In a living leaf, the carbon has to be combined with other nutrients, which are in short supply, especially nitrogen. Conserving nutrients by resorbing them into twigs during leaf senescence is a brake on the flow of nutrients between plants and soils. But the energy that remains in the carbon bonds in the dead leaves is still available to support microbes in the soil food web. Perhaps a leaf is not entirely dead until its last carbon bond is broken by microbes and the carbon is released as carbon dioxide, which drifts away to be taken up by a new leaf somewhere else, and the cycle starts anew.

PART III

Foraging, Food Webs, and Population Cycles of Predators and Their Prey

The carbon and nutrients assimilated into plant leaves and stems are next passed on to herbivores that eat the plants and in turn to predators that eat the herbivores. This network of who-eats-whom interactions is known as a food web. An organism's role in the food web is known as its trophic level. Plants at the base of a food web are level 1, herbivores are level 2, predators that eat herbivores are level 3, predators that eat those predators are level 4, and so on. An ecosystem includes the food web and the pools of nutrients in soils, sediments, or water that the plants draw upon. By analogy with trophic levels in a food web, these nutrient pools in the ecosystem can be thought of as trophic level 0. Finally, the microbes and fungi that recycle the nutrients from dead plants and animals into the nutrient pools would then be trophic level −1.

Besides the trophic level a species belongs to, the role of a species in a food web also depends on how it shuttles energy and nutrients from the organisms it eats in the trophic level below it to the species in the trophic level above that eat it. Because northern food webs consist of only a few species, they are distinguished from more southern food webs by their simple structure. But this is not to say that they don't have complex behaviors. Changes in a species' population can increase or decrease the transfer of energy and nutrients through it to the rest of the food web. Because each species in a northern food web behaves and transfers nutrients and carbon in very different ways from each of the other species, changes in one species can therefore have large effects on the other members of the food web and the cycling of energy and nutrients through the ecosystem.

The acts of traveling in search of foods and then killing, biting, chewing, and eating them are known collectively as **foraging**. An animal spends much energy foraging for food, so the way it forages and the foods it selects together determine its energy and nutritional balances. Animals that have a greater net energy and nutrient balance stand a better chance of leaving descendants. Foraging behavior is determined partly by the genes an animal inherits from its parents, so a successful forager will bequeath those genes to the next generation. This is natural selection in a nutshell.

It doesn't pay for an animal to choose foods that take more energy to procure than it obtains from the food itself. Animals must therefore select certain foods and avoid others. This is especially true for herbivores because the plant foods available to them are often not very nutritious. In the case of browsers such as moose, beaver, and deer, the woody branches and twigs they eat are not much more nutritious than a pencil. By including certain foods and avoiding others, herbivore's foraging behavior can alter the relative abundance of the foods available to it. This poses problems for the animal itself and its descendants; it also strengthens or weakens the roles of these different plant species in the cycling of energy and nutrients through the ecosystem.

Sometimes the couplings between plants and herbivores and between predators and prey generate extreme up-and-down cycles in their populations. The causes and consequences of these cycles are still not well understood. Population cycles are common in almost all northern ecosystems and rare south of the North Woods, but why this is the case is still a mystery. Population cycles of northern species lie at both the foundation and the cutting edge of population, community, and ecosystem ecology.

How should individuals forage in northern environments? Given the few types of food available to them, how should herbivores and predators forage to meet their energy and nutritional needs? How does natural selection shape these populations and the food webs composed of them? How do the decisions a foraging animal makes to meet energy and nutritional needs control the flow of energy through the northern ecosystem? Although these questions about the structure and behavior of northern food webs and ecosystems remain open to further investigation, answering them requires that we begin with a sound knowledge of the natural history of the animal, the foods it eats, and the species that eat it.

9.
Foraging in a Beaver's Pantry

A beaver must spend much energy roaming in the upland around its pond to find food and then cutting down the tree and dragging branches back to its lodge. What foods should it choose to balance the energy cost of procuring them?

Every evening in the fall, as the sun sets a minute or two earlier than yesterday, beavers enter the forest surrounding their ponds. This forest is their pantry. The beavers are searching primarily for aspen or willow but sometimes maple, birch, oak, cottonwood, or ash. During the evening and into the night, the beavers will cut, haul overland, and float branches and small stems across the pond back to the food cache a few meters beside their lodge. The beavers weigh down the branches with mud and rocks, so this food cache is mostly underwater. When winter comes, the layer of ice atop it seals it in. Protected from predators by the overlying ice, the beavers swim out from underwater entrances to their lodge and bring back small branches from the nearby food cache. After dragging the branches to platforms above the water within the safety of their lodge, the beavers then nibble the bark off them for their daily meals during the long winter. Because the beavers will not emerge from their lodge or from beneath the ice until the next spring to forage again in the surrounding forest, the food cache is their entire winter's food supply. The amount and quality of food in the cache under the ice therefore determine whether the beaver colony (usually a family unit of two parents and three to five children) survives the winter. In his classic study of the natural history of the American beaver, Lewis Morgan estimates that the food cache can range up to a full cord of wood (a cord is a stack of wood 4 feet wide by 4 feet deep by 8 feet long).[1]

Anyone who has cut, hauled, and stacked a cord of firewood for the winter knows that building a food cache this big requires a large expenditure of energy. But there are other reasons why building a food cache large enough to provide the energy and nutrition to get the beavers through the winter is difficult. First, the lodge lies in a central location in the beavers' foraging area, so beavers must travel away from their lodge to the tree they are felling and then drag the branches back. Because each trail must be traveled twice for each branch, beavers incur double energy expenditures just in traveling. Add to that the energy costs of dragging branches back through the brush, and you can appreciate the amount of energy beavers need to expend to ensure a winter's food supply. Spending too much energy to make an overly large food cache is wasteful, but spending too little and not having enough food is a disaster. Natural selection removes such individuals from the gene pool quite effectively. In addition, wood and bark are not very nutritious. How can a beaver offset the high costs of moving a cord of wood and bark to their lodge when wood and bark are poor-quality food?

This particular problem is known as the central place foraging problem because the beaver has to travel out from the lodge at the center of its home range, then bring food back to it. This is a problem faced by any animal that forages out from and back to a central location, such as bees and wasps around their hives and birds around their nests. What foods should a central place forager such as a beaver choose, and how far away should it forage to at least balance the energy costs of travel to and from the lodge? The beaver needs to maximize the ratio of the energy or nutritional content of the food in its cache to the sum of energies expended in cutting, hauling, and traveling from and back to the pond. This approach is a sort of benefit/cost accounting of building the food cache each fall. It is rather straightforward to measure the energy value of a food cache by taking a small sample of it, but it is much more difficult to measure the energy expended by the beaver in traveling, cutting, and hauling. But because energy expenditures for each of these activities are proportional to time spent in doing them, the benefit/cost ratio can be more easily measured by ecologists as the ratio of energy value in the food cache to total time spent in each of these activities during foraging.

The beaver must forage in a way that maximizes this energy gain/foraging time ratio. To do this, it can either choose foods with maximum net energy gain per energy expended in cutting and dragging the branches back to the food cache or minimize the time it spends traveling, cutting, and dragging branches back. A beaver must decide which foods to eat and which foods not to eat. This does not necessarily mean always eat this and never eat that, although those are two possibilities. There are also different degrees of selection. Intentionally selecting for or against something means choosing it or rejecting it in greater or lesser proportions than it is available. For example, suppose aspen constituted 70 percent of all the stems at a given distance from the pond. If 70 percent of the stems cut by a beaver are aspen, then the beaver is not selecting for or against it but is cutting aspen stems at random in proportion to their presence. In contrast, if aspen constituted 90 percent of the stems cut by the beaver around this pond, then we would say that the beaver is selecting for aspen because it is cutting them at a greater proportion than their availability. On the other hand, if aspen constituted only 55 percent of the stems cut around this pond, then the beaver is selecting against it, even though aspen stems are more than half of the stems cut.

Beavers can choose or avoid certain species, but they can also choose or avoid certain sizes of trees. The time it takes a beaver to cut a tree,[2] and hence the energy expended, increases exponentially with the tree's diameter. However, once they are cut, the larger trees provide an abundant source of branches that can be lopped off and dragged back to the lodge. Conversely, smaller trees are easier to cut, but it takes many small trees to provide the same amount of food as could be gathered from one large tree. So cutting larger trees seems at first to be the better strategy.

But size and species are not the only factors controlling energy gained from a tree the beaver cuts. The net energy or nutrition gained from a cut tree depends also on how far the stem or branch must be dragged back to the pond. Because the beaver is trying to maximize not only the amount of energy gained but the net energy gained per time spent foraging, the optimal diameter of a tree the beaver chooses to cut may depend on how far it is from the pond and therefore how long the beaver must travel to get to it and drag it back. One of the first predictions of central place foraging theory is that the forager should select progressively larger prey the farther the prey is encountered from the central lodge, nest, or burrow. The underlying biological reason is that it doesn't pay to gather food by carrying many small items back to the lodge in many trips instead of carrying one large piece in one trip (up to a practical limit of how much can physically be handled).

Beavers are important test cases for central place foraging theory because we can precisely measure the distance from the lodge to each tree and the size and species of each tree cut. Trees cut by beavers are easily identified by the cone-shaped stump and piles of wood chips at their base. We can then compare the data on species and sizes of cut trees and their distances from the pond with the sizes and species of all trees available at different distances from the pond. These data can then be compared with theoretical predictions.

It turns out that beavers sometimes followed the theoretical predic-

tions, but sometimes they did not. This is good not only for the beavers but also for us because things get interesting when our well-thought-out (and well-cherished) hypotheses or predictions are rejected. This is why we shouldn't be afraid to reject our most cherished hypotheses: The way nature really works is always far more interesting than what we originally thought. This is one of the more difficult and uncomfortable (but also more valuable) things we all learned in graduate school.

Although several researchers have found that beavers are indeed more size-selective with distance from the pond, there was no consistent trend in the sizes they select for and against with distance. Around some ponds, beavers cut more larger and fewer smaller trees[3] farther from the pond compared with those cut closer, as predicted by the theory. But around other ponds they did the opposite.[4] As if these conflicting findings were not confusing enough, around still other ponds beavers preferred to cut intermediate-sized trees, but with increasing distance from the pond the beavers cut fewer trees from the smaller end and more trees from the larger end of the intermediate size range.[5] Although the scientific literature may be confused, we can be sure that the beavers are not: Natural selection is very effective at ridding the gene pool of confused animals.

The direction of size selection for or against larger, smaller, or intermediate size classes must depend on other factors. The studies cited earlier examined only one or a few ponds. One way to proceed would be to get more data from a larger number of ponds: More sites, more plots, and bigger plots are often the answers to many ecological conundrums. This is what Daniel Gallant set out to do as an undergraduate at the University of Moncton in New Brunswick, Canada.[6] Gallant hypothesized that the way beavers select for different size classes with distance from the pond may depend on habitat quality. Because beavers everywhere prefer deciduous species over conifers, Gallant defined habitat quality as the proportion of stems around a pond that are deciduous species: The higher the proportion of deciduous species, the greater the habitat quality because the beavers don't have to spend as much time searching for them as when they are scarce. Gallant then surveyed beaver-cut stems around twenty-five ponds in Kouchibouguac National Park, New Brunswick. He found that in high-quality habitat beavers preferred to cut larger trees from their preferred deciduous species as they traveled farther from their lodge. In contrast, when beavers had ponds in poor-quality habitat, the size of the tree cut did not differ with distance from the lodge. In poor-quality habitat with fewer preferred deciduous species, beavers had to make do with what they could get and could not afford to be choosy about the size of trees they select at any distance from the pond.

As they cut aspen and other deciduous trees over many years, beavers can sometimes drive the forest to become dominated by the less preferred spruce, fir, and pine.[7] These conifers contain high concentrations of phenolics, tannins, and other compounds that are very bitter. Phenolics and tannins are what give coffee and unsweetened pure cocoa their bitter taste, for example. Therefore, many herbivores avoid conifers with high concentrations of these compounds. But because beavers (and, as we shall see, moose) choose mainly aspen and avoid conifers, the forest becomes more dominated by conifers over time. It is possible that beavers may become less choosy when they drive down the quality of their habitat around a pond so that the forest does not become so dominated by conifers, but I know of no long-term study that has tested this possible change in beavers' foraging behavior as the forest around their pond changes.

When beavers first build a pond and begin foraging, aspen trees of all sizes have similarly low concentrations of phenolics in their bark. At this stage, the beavers pay little attention to phenolic concentrations in the bark because these concentrations are all low and roughly the same. After the larger aspens are cut, however, their roots sprout new stems. These new stems have higher concentrations of bitter phenolic compounds compared with the larger trees of the parent generation.[8] Perhaps the resprouted aspen of the next generation are induced by the cutting to produce more bitter phenolics to defend themselves. The beavers or their descendants now begin to avoid the small, resprouted aspen stems to minimize the amount of bad-tasting food per time spent foraging. The chemical response of the aspen trees to being cut thereby altered the beavers' foraging strategy from maximizing energy gained to minimizing the amount of phenolic compounds in their diet.

The production of bitter compounds by each tree may be genetically determined. Joseph Bailey and colleagues fed branches of Fremont cottonwoods, narrow-leaved cottonwoods, and various hybrids between them to captive beavers in a cafeteria experiment where the beavers were allowed to choose which branches to eat.[9] Fremont cottonwoods produced far less tannins, another set of bitter compounds, than did narrow-leaved cottonwoods. Various hybrids between the two species had different proportions of genes that came from the Fremont parent and hence different concentrations of tannins. The more Fremont genes in a hybrid, the lower the tannin content of its bark, and the more the beavers preferred it. The plant community around a pond might therefore depend on the genes that control the chemistry of the bark and how long the beavers have occupied the pond. By discriminating between trees on the basis of their chemistry, the beavers also alter the genetic composition of the next generation of trees. Beaver foraging behavior may be a strong natural selection pressure on the distribution of genes that control plant chemistry.

It will take more research to learn whether production of phenolic compounds and tannins also depends on soil chemistry or climate and whether the chemistry of the bark changes with distance from the pond as the land rises higher from the water table. It would be interesting to follow beavers and the associated plant populations around their ponds for several generations to see whether foraging by beavers drives changes in the plant gene pools and whether the beavers subsequently alter their foraging strategies in response. In the meantime, beavers will continue to climb out of their ponds every fall and do the best they can with what the forest has to offer in order to build their food caches for the winter, just as beavers have always done.

10.
Voles, Fungi, Spruce, and Abandoned Beaver Meadows

Voles control whether conifers invade abandoned beaver meadows by dispersing the spores of mycorrhizal fungi that help the spruce seedlings take up nutrients from the soil.

The upland forests around beaver ponds are often composed of overstory aspen and understory spruce and balsam fir. These are quintessential North Woods tree species. The vertical structure of these forests is readily evident to anyone sitting in a canoe in the pond, especially in the fall, when the golden aspen crowns lie between the dark green of the conifers below and the cerulean blue of the sky overhead. In the fall, just after the blue-winged teal begin to migrate and while the geese are flying, beavers cut the overstory aspens and drag the small branches and twigs to their ponds for their winter food cache next to the lodge. The cut aspen trees sprout again from their roots, but moose sometimes browse them heavily for their high nutrient content, avoiding the spruce and fir because their needles are difficult to digest. Freed of competition from the aspens, the understory spruce and fir grow taller and become the overstory, all the while casting a dense shade and inhibiting aspen seedlings and root suckers from growing. Eventually, a dark wall of spruce and fir erects itself around the pond margin, especially around smaller ponds where the beaver can cut the entire population of aspen. Bereft of their preferred food, the beavers seek greener aspen pastures elsewhere and abandon the pond.

Without the beavers there to repair it, the dam soon fails, and the pond begins to drain. Just as the aspens in the upland are replaced by conifers, the flooded pond is replaced by a wet meadow. As the dam deteriorates during the next several decades, the water table in the meadow gradually descends. Sedges and irises and then various grasses invade the meadow, forming a thick cover reminiscent of prairies farther west in the Dakotas.[1]

But the conifers remain firmly in place on the edge of the meadow. Few, if any, conifer seedlings can be found in the meadow, even a few tens of meters from the growing dark wall of spruce and fir. The seedlings that do establish themselves last but a year or two and then expire, but inside the forest, conifer seedlings sprout and grow into saplings and eventually mature, cone-bearing trees.

Why do all conifer seedlings die within a year or two in the meadow but survive and grow into saplings in the adjacent forest? It cannot be competition between the conifer seedlings and the dense grasses, because balsam fir and spruce readily invade powerline right-of-ways and roadsides, which also usually have a dense grass cover. It cannot be that the soil in the meadow is too wet for balsam fir and spruce, as they can grow in peatlands that are often wetter than the drained meadows.

More than half a century ago, in 1950, Sergei Wilde, the dean of scientists studying forest soils,[2] suggested that conifers could not invade beaver meadows because the meadow soils lack a suite of fungi known as **mycorrhizae**.[3] These fungi, whose species number in the thousands, form a tight symbiosis with plant roots, interpenetrating the root tissues but also extending the root network further into the soil. The fungi receive carbohydrates produced by the tree's photosynthesis and in turn supply the tree with additional nutrients such as nitrogen and phosphorus, which the very fine and rootlike fungal hyphae take up from the soil. This carbohydrate-for-nutrient swap between the mycorrhizal fungus and the tree sustains both the host tree and the fungus for the lifetime of the tree. Some fungal species form associations only with specific tree species, whereas others have a broad range of host tree species they can associate with. Most trees have associations with many different fungal species. The mycorrhizal network branching off the tree's roots is a small garden of fungal diversity. Without this symbiosis with mycorrhizal fungi, most seedlings die within a year or two.

Wilde found that the mycorrhizal fungi in the meadow are killed during the long period of time that the beavers actively maintain the dam and flood the soil behind it. Wilde proposed that when the pond is abandoned and then drains, seedlings of spruce and fir cannot immediately grow in the soil in the meadow because the soil is bereft of the mycorrhizal fungi these species need to obtain nutrients.

But as with all good problems in natural history, answering one question only raises more. Obviously, there must be mycorrhizal fungi in the forest immediately adjacent to the meadow because spruce and fir seedlings grow quite well there. If this is the case, why haven't these fungi reinvaded the meadow even after a few decades of the meadow being drained? Shouldn't the spores produced by their fruiting bodies have dispersed into the meadow by now? (The fruiting body of a fungus such as a mushroom is the cap we typically eat.) After all, the soil in the meadow is no longer flooded and is sometimes nearly as dry as in the surrounding forest.

The answer to this question lies partly in the fact that most of these mycorrhizal fungal species have their fruiting bodies belowground. But how can they possibly disperse their spores throughout the forest and into the adjacent meadow? The answer to this question lies in another symbiosis, now between the fungi and the red-backed vole, which burrows in the soil and finds fruiting bodies of fungi and consumes them.[4]

Fungi are high in mineral nutrients and carbohydrates and are an excellent source of food for the voles. The spores are not digested by the vole's stomach and intestines but instead are voided whole and viable in fecal pellets. In a study of the diversity of mycorrhizal spores in red-backed vole fecal pellets in Minnesota's North Woods, we identified fifteen different species of fungi.[5] Whereas most of these species form symbioses with a broad range of hosts, including shrubs, deciduous trees, and conifers, three of the species specialize on conifers. Several of these species previously had been found only in the Pacific Northwest but not here in Minnesota. These findings are significant because they document a possible extension of the known range of these species into the Lake Superior region, a thousand miles away from the Pacific Northwest. If mycologists want to document the diversity of fungi in an area, they might do best to trap voles and examine their fecal pellets. The voles are probably much more effective at finding fungi, especially those that fruit underground, than we could ever be by sampling the soil.

Documenting the broad diversity of mycorrhizal spores in vole fecal pellets answers one question but prompts several others: Are these spores capable of forming symbioses with spruce or fir seedlings after passage through the vole's digestive system? If we inoculate beaver meadow soil with vole fecal pellets containing these spores, will spruce seedlings then be able to grow? Answering these questions was the objective of a thesis of John Terwilliger,[6] one of my graduate students.

To do his experiments, John needed a source of both spruce seedlings and red-backed vole fecal pellets. The first was easy: He obtained seeds from the Minnesota Department of Natural Resources and germinated them in sterilized potting soil (the soil had to be sterilized to ensure that he began with seedlings without any mycorrhizae). Red-backed vole fecal pellets are not available from any scientific supply company, so John simply went into the woods and caught the voles nightly with live traps. Voles have the convenient habit of voiding when they are first trapped, so every morning John collected the fecal pellets in the trap and, after marking the vole to determine population density, released it. The vole scurried away, probably a bit perplexed but unharmed. John then put out a clean trap for the next night until he had an adequate supply of vole fecal pellets, perhaps the largest supply of vole feces anyone has ever collected then or since. He trapped voles in both May (early spring here in northern Minnesota) and August (the end of summer) to see whether there was any difference in the amount and types of fungi in the voles' diet during the growing season.

John next examined the fecal pellets for spores of mycorrhizal fungi and found many of the same spores that we had found previously. He then collected soils from three beaver meadows and three adjacent forests and assigned subsamples from each of them into one of four groups: meadow soils inoculated with fecal pellets collected in May, meadow soils inoculated with fecal pellets collected in August, meadow soils that remained uninoculated to serve as a control, and the forest soils. He then grew the spruce seedlings in small pots in soils from each of the four groups in a greenhouse. (John did this research while he was also an instructor at Vermillion Community College, the academic home of Sigurd Olson, in Ely, Minnesota, an hour and a half drive north of my campus. He needed to check the seedlings and water them every day, so doing the work in greenhouses on my campus was not practical. At the time, Vermillion did not have greenhouse space for John's experiment, but the Fat Chicken Feed Store in nearby Winton generously loaned him greenhouse space. Thank you, Fat Chicken.)

John harvested some of the spruce seedlings after 12 weeks and the rest after 48 weeks, weighed and dissected them into roots and shoots, and examined their roots for mycorrhizal fungi hyphae, which are easily seen under a dissecting microscope. The results of this experiment were the sort of conclusive results we all hope for but rarely get. None of the seedlings in beaver meadow soil without vole fecal pellets had any mycorrhizae. These promptly died. In contrast, 93 percent of the seedlings grown in forest soil collected in May and 100 percent of the seedlings grown in forest soil collected in August had formed symbioses with mycorrhizae. Almost all of them survived. What's more, about 30 percent of the seedlings grown in meadow soils inoculated with vole fecal pellets formed symbioses with mycorrhizae and survived. Clearly, forest soils contain sufficient mycorrhizal spores to inoculate spruce seedlings, and meadow soils contain none, but adding vole fecal pellets to meadow soils provided a source of mycorrhizae to the seedlings. In addition, seedlings grown in forest soils weighed twice as much as those grown in meadow soils without fecal pellets, and the seedlings grown in meadow soils with fecal pellets weighed somewhere in between. The soils inoculated with August pellets produced larger seedlings than those inoculated with May pellets, possibly because the fungi had an entire growing season to produce fruiting bodies by August but only a few weeks since the end of winter in May.

The fecal pellets could also add nutrients to the soil, and we had previously found that pellets rapidly release their nitrogen and phosphorus upon decay. Thus fecal pellets could also benefit the seedlings by acting as a fertilizer. But the amount of nitrogen and phosphorus in the pellets John added was much, much less than that needed to support the growth of seedlings. So the evidence is in favor of pellets aiding seedling growth by supplying mycorrhizal fungi rather than by supplying nutrients.

Besides the differences in seedling growth between the groups, there were also interesting trends in the allocation of seedling growth to roots (to obtain water and nutrients from the soil) and shoots (to capture light to drive photosynthesis). Bereft of any mycorrhizal fungi to extend their root system with hyphae, the spruce seedlings grown in the uninoculated meadow soils had to produce more of their own roots compared with seedlings grown in the forest soil or the inoculated meadow soil. But the seedlings in uninoculated meadow soil were also smaller than those that formed associations with mycorrhizae in the forest and inoculated meadow soils. These differences in seedling growth indicate that it is more effective for a seedling to supply carbohydrates to mycorrhizal fungi to produce hyphae than to produce its own roots to capture nutrients.

Are the red-backed voles traveling from the forest into the meadow and depositing fungal spores as they go? To answer this question, John also placed live traps in the beaver meadows adjacent to the forests where he captured the red-backed voles for his supply of fecal pellets. Red-backed voles were caught in all forests, but only one was caught in a meadow, and that one was caught on a small island in the middle of the meadow on which eight black spruce trees grew. Clearly, red-backed voles stay out of beaver meadows, and so the mycorrhizal fungi population that was killed when beavers flooded the pond could not be reestablished through dispersal by the red-backed voles.

So why don't red-backed voles go out into beaver meadows? Perhaps the dense grasses are not to their liking. Red-backed voles seem to prefer forests with an abundance of down logs, whose lower sides provide cover from owls and whose tops serve as runways instead of grassy meadows without logs.[7] But there may be a more interesting reason. The most common rodent John caught in the beaver meadows was, appropriately enough, the meadow vole, which he never caught in the forest. Meadow voles and red-backed voles are very aggressive against each other and can exclude each other from their respective preferred habitats.[8] Wars fought between these two species of voles at the boundary between forest and meadow might prevent red-backed voles from entering the meadows even if they were inclined to do so.

The obvious next experiment would be to fence and enclose a large area of meadow extending partly into the adjacent forest and trap and remove all meadow voles inside. Then, the end in the forest should be opened so that red-backed voles could enter it. If the aggression by the meadow voles is the factor keeping red-backed voles out of the meadow, then eventually we should begin to trap red-backed voles inside the enclosure once meadow voles are removed. Given enough time for the red-backed voles to inoculate the meadow soil with fecal pellets, spruce or fir seedlings should become established in the enclosure. But if the red-backed vole simply prefers forested habitat, then we should never trap red-backed voles in the meadow, even if its enemy the meadow vole is removed. Without red-backed voles, mycorrhizae could not inoculate the meadow soil, and spruce and fir would be prevented from invading. We have not done this experiment, but here it is, an open invitation for an enterprising graduate student with fence-building skills. Science is not always done by expensive and complicated technology such as DNA sequencers and particle accelerators. Sometimes a simple fence and a few live traps are enough.

You can tell when a research problem is a good one when it begins to take you far afield from the question that initially motivated you. We began by asking why the boundary between a conifer forest and a beaver meadow is so sharp and permanent despite the dispersal of conifer seeds into the meadow, implying that they are entirely separate ecosystems. But we ended up learning that meadow and forest are stitched together by the foraging behaviors of beaver, moose, red-backed voles, and meadow voles.

11.

What Should a Clever Moose Eat?

What foods should a moose eat, where should it eat them, and what are the consequences?

Imagine you are a moose. It is −40°C and you are standing in snow half-way up your legs. You live on a diet of twigs about the size of a pencil and not much more nutritious. You need 6 to 9 kilograms (dry weight) of twigs each day to offset energy losses from walking, running, and just trying to stay warm. And if you are a cow, you are probably pregnant. How should you eat to stay alive? Makes our weekly grocery shopping trips look simple, doesn't it? So what should a poor moose do?

Moose are **ungulates**, the group of hoofed herbivores including other members of the deer family as well as horses, bison, cattle, sheep, and goats. Owen-Smith and Novellie suggest in a famous article (from which I cribbed the title of this essay) that in the short term a clever ungulate can maximize its energy or nutrient intake per unit time spent foraging by choosing the most nutritious of all the foods available and eating as much of them as it can without moving to a new spot.[1] For a moose in winter, none of the twigs available are especially high-quality food. The first thing a moose must do when faced with generally bad food is to eat the best of the bad. In winter, these are twigs from the more nutritious deciduous species such as aspen, birch, and willow.[2]

Like all ungulates, moose digest their food not so much by the acids they make in their stomach as by bacteria living in their rumen. The **rumen** is a pouch that the food is first deposited in after the moose swallows it. The microbes in the rumen ferment the food. The food is then regurgitated into the moose's mouth, chewed again, and then swallowed again, whereupon the food enters the stomach proper. Aspen, birch, and willow leaves and twigs are easier for the rumen bacteria to digest than conifer needles because of their low lignin and cellulose contents. In contrast, if moose eat twigs of balsam fir or spruce, the high lignin and cellulose contents in these twigs will give them a bellyache. This is because lignin and cellulose in spruce and fir needles are difficult for rumen microbes to digest. In addition, the high phenolic content of the needles will probably cause the renal tubes in their kidneys to bleed. In late winter, when moose have browsed most aspen, birch, and willow above the snowline and are forced to eat balsam fir, it is not uncommon to find red urine spots in the snow along a line of moose tracks.

But restricting the diet to small aspen, birch, and willow poses other problems to moose, such as finding these species. Because aspen, birch, and willow within reach of moose are intolerant of shade, they need patches of full sunlight to maintain growth. When taller trees shade them, they cannot maintain high enough photosynthetic rates to survive. The supply of these open patches, and therefore the populations of aspen, birch, and willow, has to be maintained by periodic disturbances that remove the overstory, which can range from small and frequent patches of cutting by beavers around their ponds to infrequent crown fires that kill overstory trees across large chunks of the landscape.

In the North Woods, clearings around beaver ponds are the most reliable source of aspen, birch, and willow. After beavers fell large aspen around their ponds for their own food supply, the aspen roots sprout many thousands of sucker stems within reach of moose. Shoots of willow and birch also benefit from the increased light in beaver clearings. There are numerous beaver clearings in many valley bottoms, but they are small and don't provide much food. A moose must therefore make a circuit around the landscape to visit a number of beaver openings to obtain sufficient food.

In contrast to beaver openings, fires provide abundant food because they are large and are often colonized by aspen and birch soon after. If it finds a recent burn with abundant forage, a moose need not travel as much to find food as it does if it needs to rely solely on beaver openings. But fires are less frequent and less numerous than beaver clearings, and many moose can spend their entire lifetime without encountering a burn in their home range.

Val Geist proposes that natural selection has produced two different reproductive strategies in moose,[3] each of which maximizes the number of calves against the abundance of food in small but frequent or large but infrequent disturbances, respectively. Geist proposes that cows that forage in fire-dominated landscapes should devote the abundant food available to them to two small twins. The abundant food would enable the cow to produce enough milk to supply twins, thus enhancing her reproductive output. But if the cow put the abundant food into conceiving a single large calf instead of two smaller calves, the large calf might be too large to get through the pelvic channel. Two small twins therefore stand a better chance of being born than one large calf. But eventually, the aspens grow out of the 2- or 3-meter vertical reach of the cow, conifers reinvade the burn, and the abundance of food within reach of the cow declines.

Now, the food supply for cows is maintained by the smaller but more frequent beaver openings. Cows giving birth to twins would now be at a disadvantage because the smaller beaver openings may not supply enough food for the cow to provide sufficient milk to sustain two calves.

Cows that conceive a single and viable but not large calf are now at an advantage when aspen is supplied by beaver clearings compared with cows that produced twins when aspen was regenerated in abundance after large fires. Therefore, disturbance size and frequency are two forces selecting for one or the other reproductive strategy of moose. I know of no direct test of this interesting idea, but perhaps measurements of moose calf numbers and sizes in different landscapes with different fire frequencies and beaver population densities might provide some observational evidence for or against it.

If moose choose to browse aspen and birch and avoid most conifers, can moose increase the abundance of unbrowsed conifers? In the late 1940s, Laurits Krefting of the University of Minnesota addressed this question by building four moose exclosures on Isle Royale.[4] An exclosure is a fenced area that keeps out, or excludes, a particular animal from the plant community inside it. Inside the exclosure, the effect of an animal on the plant community is effectively shut off. The exclosures on Isle Royale are fences 12 feet high and approximately 100 feet to a side that protect the forest inside from moose browsing. These are now the oldest animal exclosures in continuous existence anywhere in North America and possibly the world. Because we've been able to observe how the forest changes in the absence of many generations of moose, these exclosures provide valuable and rare evidence of how herbivores such as moose control plant communities and ecosystems through their foraging decisions.

Protected from moose by the exclosures, the aspen and birch inside them began to grow rapidly and eventually became a continuous overstory canopy. Outside the exclosures, heavy moose browsing caused aspen and birch to maintain a shrubby, stunted form and eventually die. The forest outside the exclosure is now dominated almost entirely by spruce, which the moose refuse to touch. The few large aspen and birch that can be seen outside the exclosures almost always date from the late 1930s, which was a period of low moose population density on Isle Royale.

This difference in vegetation inside and outside the exclosures suggests that moose might even be changing soil fertility and the cycling of nutrients. Outside the exclosures, less leaf litter is returned to the soil, and that leaf litter is often dominated by spruce needles.[5] Like the bacteria in the moose's digestive tract, the bacteria in the soil also decompose leaf litter, releasing nutrients needed for plant growth. Just as spruce needles are not easily digested by the bacteria in a moose's stomach, they are not easily decomposed by the bacteria in the soil. The slow decomposition of needles decreases the rate at which nutrients are recycled for plant uptake. This suggests that the soil outside the exclosures, which received mainly spruce and fir needles from unbrowsed trees, should be less fertile than soils inside the exclosures, which have received decades of more nutrient-rich and more rapidly decomposing aspen and birch leaf litter. My technician Brad Dewey and I tested this hypothesis by measuring nitrogen availability inside and outside these exclosures 40 years after Krefting built them. We found that soil nitrogen availability was as much as 50 percent lower outside the exclosures compared with inside.[6] This difference in nitrogen availability inside and outside the exclosures meant that the productivity of plant biomass was as much as 30 percent lower outside the exclosures compared with inside.

The size of this difference amazed us: It meant that, despite their low population density, a solitary animal such as a moose could have a large effect on the cycling of nutrients through an ecosystem. Previously, it had been thought that only animals that foraged in large herds, such as bison or wildebeest, could alter plant productivity and nutrient cycling.[7] In contrast to moose, these herds of ungulates increased nutrient cycling and productivity by manuring the soil with fecal material and urine.

So why doesn't moose manure outside the exclosure partly counter the decline in soil fertility compared with inside the exclosure, much like the cow manure my grandfather used to fertilize his fields? After all, not only did the exclosures stop the moose from browsing on aspen and birch, it also stopped them from depositing nutrients in manure. To examine this further, we needed to collect some fresh moose droppings and see how fast they released nitrogen. Fortunately, one day we came upon a fresh, steaming pile of fecal pellets. We collected some of these pellets and incubated them in the laboratory under optimal temperature and moisture conditions for bacterial and fungal growth and measured the rate at which nitrogen was released from them. We also incubated soil we collected from an island off of Isle Royale, which, being small, lacked a resident moose population. The moose pellets had lower rates of release of nitrogen than the soil from the island without moose. The reason for the low rates of nitrogen release from moose pellets may be that the moose removes most of the nitrogen from the twigs during digestion and excretes only the remaining undigestible plant material, which now has the consistency and chemistry of sawdust. Moose pellets are not good manure for your garden.

Moose may be clever ungulates in the short term because they browse the most nutritious aspen and other hardwoods, as Owen-Smith and Novellie suggested. But over one or two generations, this strategy appears to be not so clever because continuous browsing on shade-intolerant aspen, birch, and willow causes them to be overtopped by less preferred spruce and fir, which in turn causes productivity and soil fertility to decline. But if the ability of the ecosystem to sustain moose populations is declining because of moose browsing, then why are moose not driving themselves to extinction? Clearly, we are missing something.

Over the years, we noticed that aspen, birch, and willow stems that are lightly to moderately browsed seem to become bushier. Where do the new branches come from? Aspen, birch, and willow have small buds beneath their bark, known as **epicormic buds**. These buds are normally quiescent, being suppressed by the hormone auxin, which is produced by the terminal bud at the end of the twig. Auxin is a sort of "master hormone" that controls many aspects of plant growth, such as promoting the growth of the shoots at the ends of branches while simultaneously suppressing the growth of side shoots. When a moose browses the apical bud or portion of a twig, the auxin is no longer produced and therefore no longer suppresses the epicormic buds lower down in the twig. These buds then sprout into two side shoots. Now there are two twigs for a moose to browse instead of one main twig. Perhaps a moderate amount of moose browsing increases food supply by making more twigs. But heavy browsing on these stems seemed to make them less branched and more spindly and weak.

On the other hand, when fir and pine are browsed, they often don't seem to replace their side branches with new ones. In contrast to aspen, birch, and willow, conifers have far fewer epicormic buds than aspen and birch, so far fewer new twigs are produced than in the hardwoods.

Could these differences between the way hardwoods and conifers respond to moose browsing help explain not only what we saw happening inside and outside the exclosures on Isle Royale but also how moose can persist in the ecosystem even though total productivity and soil fertility were declining? To answer this question, Nathan De Jager, a graduate student of mine, took advantage of an experiment my colleague Kjell Danell, his colleagues and students, and I began in Sweden. Kjell and his colleagues and students constructed large moose exclosures on a series of sites across a range of soil fertilities.[8] But unlike the exclosures on Isle Royale, the purpose of these exclosures was not simply to exclude moose but to allow us to experimentally mimic what different moose population densities might do to the vegetation and the soils. To so this, we divided the inside of each exclosure into four quadrats. Within each quadrat, Kjell's technicians and students clipped vegetation to simulate the browsing by 10, 30, or 50 moose per 1,000 hectares (a hectare is 100 meters on a side, about 2.5 acres). One quadrat was left as a control and was not clipped. The clipping in each quadrat was distributed between birch, Scots pine, and other species based on data Kjell had collected over the years on the diet composition of moose. Moose pellets from a nearby game farm and artificial urine were added in proportion to the simulated population density.

After several years of clipping, the branching structure of birch and Scots pine crowns began to change, and we needed to measure these changes somehow. De Jager did this by first constructing a sampling frame he could place over the crown. Two sides of the frame were partitioned into strips 10 centimeters wide with string. By looking through each strip and counting the number of twigs and measuring their length and diameter, De Jager could measure how the density of forage varied as a moose swung its head from side to side while browsing a particular sapling. From these data he could calculate the **fractal dimension** of the crown, which is a mathematical measure of its "branchiness," or the rate at which branches sequentially divide into two new branches going outward from the main trunk. Crowns that were "branchier" had more branching points and therefore more twigs at the outer edge of the crown that moose could browse. De Jager also estimated the weight of each twig from its length and diameter. From the fractal dimension and the weight of each twig, he could calculate the total amount of forage that would have been available to a moose browsing the plant. By doing this in the different quadrats in which different moose densities were experimentally simulated, he could relate the changes in forage production to our simulated moose population density.[9] And by doing this in exclosures on different soils, he could determine how soil fertility modified the response of birch and Scots pine crowns to browsing.

De Jager found that merely 4 years of experimental clipping that simulates moose browsing changes the branching structure of the birch and pine crowns as well as the twig sizes, but in different ways. As long as browsing on birch remained at low to moderate levels, the crown became branchier, which means more twigs for a moose to browse, and the new twigs were often larger than the original browsed twigs. Greater branch density and twig size on the birches meant more food for a moose. The production of forage by birch was maximized when moose browsed between 15 and 20 percent of twigs on any individual sapling. But greater amounts of browsing weakened the birch sapling, which then produced fewer and smaller twigs, so total food production declined rapidly at higher clipping rates and therefore moose population densities. In contrast, browsing reduced the pine's branchiness and twig production, but not as steeply as it did for birch at high browsing rates. So eventually, the pines grew taller and overtopped the birch.

De Jager later confirmed this finding with the same measurements on aspen and balsam fir on Isle Royale.[10] Aspen and fir responded in the same ways as the Swedish birch and pine, respectively: Aspen that had been lightly to moderately browsed had more and larger twigs than unbrowsed aspen. Like the Swedish birch, aspen produced the most forage when 15 to 20 percent of its twigs were browsed. However, heavily browsed aspen produced only a few small twigs. Browsing reduced balsam fir branch density and twig size, much like Scots pine in Sweden, but not as rapidly as happened in heavily browsed aspen. So the responses of deciduous hardwoods (birch and aspen) preferred by moose and that of the unpreferred conifers (Scots pine and balsam fir) seem to be the same in the North Woods of Isle Royale and in Sweden.

We had previously found that moose most commonly browse 15 to 20 percent of the twigs of aspen and birch on Isle Royale.[11] De Jager's finding that the maximal twig production of aspen and birch happens when moose browse 15 to 20 percent of twigs seems to explain this. When browsing was greater than this, spruce and fir overtopped the heavily browsed and weakened aspen and birch, and their forage production declined sharply. As the spruce and fir took over in heavily browsed areas, soil fertility also declined, just as we saw in the exclosure experiment.

Is the moose somehow continuously measuring the condition of its home range, forecasting what its effects might be, and adjusting its foraging strategy accordingly? This possible explanation implies that the moose are consciously altering their foraging behavior with a long-term goal in mind, namely the maximum production of forage in the long run. Such goal-seeking explanations are generally frowned upon by evolutionary biologists because natural selection operates in the short run, not to obtain long-term goals that an animal seems to set for itself. For example, if the cow moose's ability to produce calves this year depends partly on how much food the cow has eaten, why should she leave 80 percent or more of the twigs on the plant unbrowsed? Why not eat most or even all of them, especially because another moose may come along tomorrow and eat them instead?

This is a good argument, but that is not what we found the moose are doing. We still do not completely understand why the moose are not behaving as theory says they should. In Essay 9, *Foraging in a Beaver's Pantry*, we also saw that beaver did not always forage according to theory. Here are two examples where natural history observations compel us to take our theories lightly, no matter how logical and compelling they seem to us. Animals do not care about theories; they only want to stay alive. But in behaving the way they do, animals are teaching us something, if we choose to listen.

Part of the answer to this conundrum may be that moose bequeath not only their genes to the next generation but also the landscape they created. If eating most of the twigs on each aspen or birch causes forage production and soil fertility to decline within its lifetime of 10 to 15 years, then the moose bequeaths not only the genes that control that foraging behavior to its descendants but also a landscape less able to deliver the energy and nutrients its descendants need. Based on De Jager's experiments and our measurements of changes in soil fertility inside and outside the exclosures, the decline in the condition of the landscape can happen within one or two generations of moose. Such a strategy stands a good chance of being selected against in the long run because by the time the moose's grandchildren arrive, they will be faced with a less nutritious landscape with less forage. Consequently, they will be less able to produce great-grandchildren and thus propagate their ancestors' genes. If these grandchildren maintained such a counterproductive foraging strategy, they might worsen the condition of the landscape for their children and grandchildren. Within four generations of moose, the landscape may not be able to support a population of moose who eat most of the twigs of aspen and birch saplings when they encounter them.

On the other hand, moose that browse only 15 to 20 percent of twigs on any single individual aspen or birch would leave a landscape not much worse and perhaps even more productive for their progeny. Ron Moen, another of my graduate students, confirmed this prediction using a model that kept track of the energy budget of a foraging cow moose, the calves it produced, and the growth of conifers and hardwoods across the landscape.[12] This model also showed that if moose forage strictly according to Owen-Smith and Novellie's short-term optimal strategy, the preferred aspen is eventually overbrowsed, and the landscape does not provide sufficient food to maintain a positive energy balance for the cow and her calf, leading to the death of both. Of course, these thoughts about the natural selection of moose foraging behavior presume a tight correspondence between some suite of genes and the moose's foraging behavior, but at present we don't know what proportion of foraging behavior is inherited or learned.

When the behavior of moose has so much control over the productivity and cycling of nutrients, evolutionary fitness may depend as much on the landscape as on the genes they leave to their children and grandchildren. We now know a lot about how a clever moose should eat at any single instant, which is the question Owen-Smith and Novellie asked more than 30 years ago. What we have learned about moose foraging in the past several decades suggests that evolutionary biology, behavioral ecology, and ecosystem ecology might combine to address the questions: How should a clever moose forage in landscapes that previous generations of its relatives have created? How should a clever moose forage to maximize the chance that it bequeaths both its genes and a sustainable landscape for its descendants?

12

Tent Caterpillars, Aspens, and the Regulation of Food Webs

The coevolution of aspens, tent caterpillars, and their predators regulates the productivity of much of the North Woods.

Every 10 or 15 years, the North Woods experiences one of its most spectacular population cycles, the outbreak of the forest tent caterpillar,[1] or "army worm" as most people call them. But "spectacular" would not be what comes to mind for most people during these outbreaks. The most common comment would be something like "Yuck! The army worms are back! I hate those things!" At this point, the Minnesota Department of Natural Resources and the U.S. Forest Service get irate calls along the lines of "When are you going to spray these disgusting bugs?" You would think that we are Pharaoh and his people enduring a plague of locusts covering the land.

Our culture has a gut-wrenching aversion to insect outbreaks. Perhaps this is because civilization began with the invention of agriculture, which was, until recently, helpless against epidemics of locusts, grasshoppers, and other insects. Crops made the food supply more reliable, but famine usually followed when the locusts consumed the crops, so we think any insect outbreak is bad anywhere it happens. Although we don't obtain food from aspen stands, we still emotionally feel that we must save the forest from its "enemies." But as apocalyptic as these natural outbreaks of tent caterpillars may seem, the forests have always managed quite well without our help.

The forest tent caterpillar, a native species, defoliates large blocks of aspen in late spring or early summer, after the emerging leaves are well on their way to filling out the canopy, leaving them gray and leafless, as if November had made a premature return. If you walked into one of these defoliated blocks, which can be several kilometers long, you would see hundreds of caterpillars arrayed on the trunk of virtually every aspen tree. Despite their bad rap, these are really beautiful caterpillars, about two inches long and dark chocolate brown, with white spots and yellow and iridescent blue stripes along their sides.

If you listened closely while you were in this stand, you might hear a sound like the beginnings of a light rain, even if the sun is shining. It is in fact raining, not drops of water but instead a continuous outpouring of manure droppings from these caterpillars as they eat aspen leaves to fuel their transformation into moths. This "rain" of caterpillar fecal material onto anyone walking through the forest is one of the many reasons most people find army worms gross and disgusting.

By July, the defoliated aspens often produce a second crop of leaves. In general, aspen, as well as most other deciduous species, can withstand defoliation without dying (conifers, on the other hand, are almost always killed after losing their needles; see the next essay). By this time the caterpillars will have metamorphosed into tan moths, and eggs for next year's caterpillars are being laid.

The pattern of defoliation begins in one stand—even on a single tree—and grows amoeba-like across the aspen-dominated landscape, a huge blob of caterpillars spreading here and there from one stand of aspen to another like a band of desert nomads traveling from oasis to oasis. That caterpillars can mass and travel together as a coordinated group suggests that they have a social order. A social order in turn implies some way to communicate with one another. T. D. Fitzgerald and co-workers have found that tent caterpillars communicate by marking their trails with chemicals that can be sensed by all sibling members of a wandering band.[2] These trails keep the band together, allowing it to regroup after feeding bouts in large masses that Fitzgerald calls "bivouacs," like platoons of some vast army, hence the name *army worm*. This mass movement of tent caterpillars is another trait people find disgusting, especially when roads are sticky with thousands of their mashed bodies. Sometimes, snowplows are taken out of their summer hibernation to clear the roads.

It is not clear what the advantage of staying together in large masses is, although keeping each other warm during cold spring nights is one possibility. Another possibility is called predator saturation and depends on the hope that if there are enough of you and your relatives together in any one spot, your predators could not eat all of you, and therefore some of your siblings who share some of your genes will survive, reproduce, and perpetuate your genetic code in future generations.

Trees do not sit passively back, however, and let themselves be munched by hordes of caterpillars. Instead, they have several ways of defending themselves. One defense is to increase production of various noxious compounds once caterpillars start feeding, a process known as induced chemical defense. These compounds either deter feeding by caterpillars or actively disrupt their metabolism and kill them. They include tannins, which make the leaves bitter, just as the tannins in coffee make it bitter, as well as **glycosides**, which disrupt nerve transmission, thin blood, and prevent clotting (the popular medicine coumadin is a member of this family). Glycosides are especially nasty to insect physiology.

Producing glycosides and tannins diverts a lot of the tree's energy from growth and seed production, so they are expensive for the tree to make and keep around during the years when there are few caterpillars. Perhaps the trees are stimulated to make a lot of these costly defense compounds only when caterpillars begin feeding. Otherwise, when caterpillars are not around the trees use the energy that would go into making these compounds for growth or seed production. Michael Stevens and Richard Lindroth tested this hypothesis by placing whole aspen leaves in bags with or without tent caterpillars. After a few days, they measured tannin and glycoside levels in the leaves in both types of bags.[3] If tannins and glycosides are produced by leaves in high quantities only when caterpillars feed on them, then the leaves in the bags with caterpillars should have higher concentrations of these compounds than the leaves without caterpillars. Indeed, the leaves that were partly eaten by caterpillars produced more tannins and glycosides than the leaves in the bags without caterpillars. Moreover, the bitter tannins were produced first, and production of the blood-thinning and nerve-inhibiting glycosides was delayed. Perhaps the trees first try to deter the caterpillars from feeding by producing bitter tannins, but if that doesn't work, then the heavy artillery glycosides are called into action.

Stevens and Lindroth also wondered whether soil fertility could also determine whether a tree produces chemical defenses. They reasoned that low soil fertility might not provide enough nutrients to enable defoliated trees to produce a second crop of leaves. If so, then trees on infertile soils with high amounts of tannins and glycosides before caterpillars begin feeding may be able to deter the caterpillars from feeding on the first crop of leaves and might not have to grow a second crop. On the other hand, fertile soils might provide trees with enough nutrients to grow a second crop of leaves. These trees might be better off sacrificing a crop of leaves rather than converting valuable carbohydrates to the energy-rich tannins and glycosides. To test this hypothesis, Stevens and Lindroth measured tannin and glycoside contents in leaves of trees growing on different soils in a year without a tent caterpillar outbreak. They found that aspens growing on infertile soils did indeed produce more of these chemical defenses than aspens growing on fertile soils. The trees growing on infertile soils were taking out an insurance policy, so to speak, by producing tannins and glycosides to deter the caterpillars as soon as they arrived, whereas the trees growing on fertile soils were counting on having sufficient nutrients to produce a second crop of leaves if caterpillars consumed the first crop.

The production of tannins and glycosides must be under the control of several genes that can be passed down to their descendants. Trees that can better withstand caterpillar attacks by producing tannins and glycosides when needed therefore stand a better chance of producing descendants who will inherit those genes. This is Darwin's theory of natural selection in a nutshell: The caterpillars induce chemical defenses in some aspen, but the trees that defend themselves stand a better chance of leaving descendants. In this way, the aspen population evolves with the genes to induce tannin and glycoside production.

But the caterpillars might have some genes that allow them to tolerate tannins and glycosides as well. Their descendants that inherit those genes will be able to feed on aspen despite the higher concentrations of tannins and glycosides in the leaves, but other caterpillars without those genes will not be able to survive as well. The caterpillar population will then evolve toward individuals that are somewhat immune to the tannins and glycosides. The aspens that produce even greater concentrations of tannins and glycosides that might not be tolerated by even these caterpillars are now at an advantage over the other trees. This positive feedback between two interacting species is known as **coevolution** because the evolution of each species spurs the evolution of the other. When the production of chemical compounds by a plant host is what spurs the coevolution with its insect herbivores, the process is more precisely called phytochemical coevolution.[4] The plant hosts and their insects are now locked in a coevolutionary "arms race": Increased chemical defenses by the plant spur tolerance of these defenses in the insect population, which in turn means that individual plants that produce even higher concentrations of chemical defenses are at a competitive advantage, and so forth. Although tent caterpillars and aspen would seem to be excellent candidates for phytochemical coevolution, apparently there have not yet been any genetic or ecological tests of this hypothesis in these species.

Another way that plants might defend themselves from being eaten by insects is to attract natural predators of those organisms. David Tilman found that black cherry trees do exactly this to defend themselves against eastern tent caterpillars, a species that is closely related to forest tent caterpillars.[5] The cherry trees produce organs that secrete nectar, known as **nectaries**, on the underside of their leaves in addition to the nectaries in their flowers. These foliar nectaries associated with leaves are produced during the first 3 weeks in spring, soon after the leaves emerge, exactly the time when eastern tent caterpillars are beginning their growth. While the nectaries in the cherry blossoms are attracting bees and other pollinating insects, the foliar nectaries are attracting ants that prey on the small caterpillars. The survivorship of tent caterpillars decreases the closer they are to these ant colonies. The production of these nectaries in cherries is induced by defoliation,[6] just as the chemical defenses are in aspen. It must be metabolically expensive for a tree to produce hundreds of thousands, if not millions, of extra nectaries filled with energy-rich sugar compounds, but the benefits of protecting the leaves from caterpillars by feeding nectar to the predatory ants seems to outweigh the cost of producing the foliar nectaries. This is an excellent example of a coevolutionary symbiosis between ants and cherries, each benefiting the other.

Aspens also produce nectaries on their leaves. Moreover, aspens that produce more foliar nectaries to attract ants also produce more phenolics to repel tent caterpillars.[7] Induction of both chemical and nectary lines of defense increases growth and survival of the aspen more than induction of each alone, so trees with both sets of genes stand a better chance of passing them on to future generations than trees with only one set of genes. Darwin is vindicated once again.

By attracting natural predators or producing bad-tasting chemicals, individual trees thereby regulate insect attack to some extent. Are there ways that the entire forest ecosystem regulates itself after insect attacks? Some researchers have suggested that animals may play a role in regulating forest growth, but this idea is controversial. Using computer models constructed from measurements of tree growth made by the U.S. Forest Service, William Mattson and Norton Addy hypothesize that when the tent caterpillars defoliate the canopy, the amount of light reaching understory trees increases. In addition, the conversion of leaves into nutrient-rich caterpillar manure could temporarily increase soil fertility.[8] The increased light and nutrient levels then stimulate the growth of understory balsam firs, hazels, asters, and other species during the first 2 or 3 years after defoliation of the aspen. These 2 or 3 years of defoliation every 10 or 15 years between outbreaks make up more than 20 percent of the lifespan of the aspen stand. This period of increased light penetration to the forest floor might keep these forests more diverse than they would otherwise be. If Mattson and Addy are correct, then these insect outbreaks are the ecosystem's way of regulating itself, periodically giving soil fertility and minor species a boost. However, no one has yet examined the changes in soils before, during, and after tent caterpillar outbreaks to see how much soil fertility may have increased. This is an open question waiting for the right graduate student in the right place at the right time needing a thesis problem.

Of course, very large outbreaks of tent caterpillars would not be possible without the abundance of their favorite food, namely the large stands of aspens in our northern forests. These large stands of aspens are here because of the widespread harvesting of white pines in the late 1800s that marked the beginnings of the timber industry. The white pine harvest and subsequent fires let full sunlight down onto the forest floor and in some cases stimulated soil fertility by depositing large amounts of ash, which, like the lime farmers apply to their fields, decreases soil acidity and boosts decomposition and hence nutrient recycling from dead plant litter. Full sunlight and higher soil fertility in turn increase the germination and growth of aspen and birch seeds. Being light and windblown, the seeds disperse easily across the landscape, seeking open areas in which to germinate. The large infestations of tent caterpillar are responses of the landscape to the infestation of aspens, which was an earlier response to the invasion of the forests by people and oxen harvesting the white pines. The North Woods are composed, among other things, of loggers, aspens, tent caterpillars and ants, and white pines and soil, all continually adjusting to each other.

13.

Predatory Warblers and the Control of Spruce Budworm in Conifer Canopies

The North Woods support twenty-nine species of warblers, the highest diversity of a group of related birds north of Central America. By foraging in different parts of a spruce or fir canopy, some of these warblers can partly control spruce budworm populations.

Asked where the greatest diversity of bird life is, most people would say the tropical forests, and they would be correct. Tiny Costa Rica, for example—less than a fourth the size of Minnesota—has more than fifty species of hummingbirds alone. Traveling northward, the diversity of bird life thins, until we reach the North Woods, which has the greatest number of breeding bird species anywhere north of Mexico, on average sixty to sixty-seven species per 5-mile survey route according to the Breeding Bird Survey. The family Parulidae, the warblers, is a major part of this diversity. According to my *Peterson Field Guide*, there are forty species of warblers in eastern North America; fully twenty-nine of these breed in the North Woods, three quarters of all the warbler species of eastern North America.

Despite their name, warblers do not warble. According to Peterson, warblers "zeeeeeeee-up" (northern parula), "tee-ew, tew, tew, tew" (yellow-throated), "zoozee" (black-throated green), "weesee" (black-

and-white), "weeta-weeta-weetsee" (magnolia), "zi-zi-zi" (blackpoll), "teesta-teesta-zizizi" (blackburnian), or "zee-zee-zee-zwee" (redstart). The chestnut-sided warbler is "please-please-pleased-to-meet-cha," the ovenbird is proud to be a "teacher-teacher-TEACHER," and the black-throated blue warbler asks for "beer-beer-beer."

Male warblers are extremely colorful, especially during the breeding season. Warbler feathers and even stuffed warblers used to adorn women's hats during Victorian days (fortunately, we'd rather watch live birds than wear dead ones these days). Beg, borrow, or better yet buy a sixth edition of the *Peterson Field Guide* and feast your eyes on the colors displayed in the eleven pages of beautiful paintings of male spring warblers. Yellows abound, but there are also some deep and soothing blues and many variations on black-and-white patterns, often with conspicuous yellow spots (yellow-rumped and magnolia), chestnut sides (chestnut-sided, of course), or shockingly bright orange patches (blackburnian and redstart). The sight of the blackburnian warbler or redstart is truly enough to take your breath away. During the southward migration in the fall to their tropical winter homes, most warblers lose their bright markings and become what Peterson calls "Confusing Fall Warblers" but what every birder calls Confused Fall Warblers.

Although you can see or hear these wonderful birds during most of the summer in the North Woods, the best time is during a narrow window of about 2 weeks in early or mid-May just as the leaves are emerging from their buds. April storms (extending even into early May) keep these migrants bunched south of the North Woods. Then one morning it dawns cold and clear, the type of day that makes you wonder if winter is really over, and the warblers arrive suddenly and in all their glory. Where before you could barely see three species in a day, now you can easily see more than a dozen in a few hours in the same patch of woods you scoured before without success. If you are lucky, you may be caught in the midst of a huge mixed flock known as "warbler wave" that simply swarms over and around you.

Groups of warblers form what are known as **guilds of species**.[1] A guild is a group of species that exploits the same set of resources in a similar way. If you find one species of a guild, look around for another. This will rapidly increase your bird list. All member-species in a guild need not be taxonomically related. For example, two birds that forage for insects on the forest floor, the ovenbird (a warbler) and the brown thrasher (a thrush), are member-species of the same guild but are not related. Guilds present an interesting problem in natural history: If the species in a guild each exploit the same set of resources, how do they manage to coexist and keep from driving each other to extinction?

Robert MacArthur asked this question more than 50 years ago and answered it by studying a guild of five warblers that hunt and feed on insects in the crowns of spruce trees in Maine.[2] These are the Cape May, the yellow-rumped, the black-throated green, the blackburnian, and the bay-breasted warblers. The study is so famous that this guild is now known as MacArthur's warblers. By simply sitting and watching, MacArthur found that these warblers foraged at different canopy levels or zones. The Cape May forages highest in the canopy and on the outside of spruces. The foraging territory of the black-throated green warbler overlaps that of the blackburnian, but the black-throated green forages just below and to the inside of the Cape May. The blackburnian also forages high but extends its foraging territory down farther along the sides of the conical canopies than the Cape May. The bay-breasted warbler forages below these three warblers, in the upper part of the lower third of spruce crowns, mainly in the interior of the crown. Finally, the yellow-rumped warbler occupies the bottom of the spruce canopy down to the shrubbery just over the forest floor. So, by foraging at different heights and depths within the canopy, the species in this guild not only minimize competition but also expose themselves to different kinds of insect food.

Because of the conical shape of conifer crowns, the volumes of these foraging zones grow larger as one proceeds from the Cape May warbler's zone at the top of a tree's canopy down to the yellow-rumped warbler's zone at the base. Does the yellow-rumped warbler have access to more insects because its zone is larger than the Cape May's? Or does the density of insects decrease downward so that, although each lower zone is larger than the ones above it, the total insect biomass of each zone remains roughly the same for each species? How do differences in insect density, to the extent that they exist at all, affect the search behavior and capture success by each warbler? No one seems to know the answers to these questions, but the answers are essential to a fuller understanding of the energy budget and metabolism of the warblers and their ability to feed and raise young.

Because their foraging zones together completely fill the volume of a spruce or fir tree's canopy, MacArthur's warblers can sometimes exert strong control over foliage-eating insects. In particular, the Cape May, bay-breasted, and black-throated green warblers of MacArthur's guild are major predators on spruce budworm, a major defoliator. Although it consumes both spruce and fir needles alike, spruce budworm actually prefers balsam fir (the importance of this will become apparent in a moment). Complete defoliation by the budworm kills both spruce and fir trees because neither of them can regrow needles or twigs in the same year as they are consumed.

A pair of adult warblers with five nestlings can consume 35,000 budworms during a 25-day period when the larvae are available.[3] During times of low to moderate budworm densities, this predation rate by warblers is one of the stronger controls over spruce budworm populations.[4]

Before World War II, outbreaks happened somewhere every 30 to 60 years or so, usually in mature stands with complex canopies, and lasted for 7 to 10 years. These outbreaks were usually very intense but confined to a few mature spruce–fir stands in the landscape at any one time, with the rest of the landscape harboring very low populations of budworm, which the warblers were able to keep in check.[5]

What triggers the budworm population to grow so rapidly that it outstrips the ability of warblers to control it? For many years, drought was thought to trigger outbreaks.[6] It is easy to see why drought could stress trees to the point where they could not withstand insect attack, but why should drought cause the insect population to expand in the first place? One reason may be that when trees are under drought stress, the stomates in the leaves close, and water loss through transpiration is reduced. Because transpirational loss is accompanied by evaporative cooling, the temperature in the canopy rises as transpiration declines. The warmer and drier canopy may provide a better environment for the growth of budworm larvae.[7] In addition, because trees cannot effectively convert sugars from photosynthesis to starches when they are under drought stress, the sugar content of their foliage may increase and provide higher-quality food for insect herbivores.[8] But there is little evidence that eastern spruce budworm outbreaks were caused by improvements in food quality during droughts, although this may be the case in western forests in the Rocky Mountains.[9] Droughts and budworm outbreaks may act independently to kill trees, and when they both coincide, large areas of spruce–fir forests may be killed.[10]

Gordon Baskerville, at one time dean of the College of Forestry at the University of New Brunswick, has suggested that budworms and spruce–fir forests form an interdependent, self-regulating system, and outbreaks of budworm do not need any external trigger such as drought. Instead, Baskerville proposed that budworm outbreaks are a response to the maturing of the forest itself.[11] Many spruce–fir stands have regenerated from a cohort of balsam fir seedlings that had their start in a burst of sunlight after a disturbance such as logging, fire, or even a previous budworm outbreak killed the overstory.[12] The budworm population from the seedling stage up through stands that are about 60 years old is low and effectively controlled by the warblers. As this stand matures further, the trees, now older than 60 years, form a dense and deep canopy with plenty of food for budworms. The large volume of foliage in the dense and mature canopy also partly screens many of the budworm larvae from their warbler predators.[13] By providing both their preferred food and a screen hiding them from the warblers, a large, mature, and dense crown of balsam fir provides effective conditions for a budworm outbreak to begin. Now, the number of budworms is too great for the population of warblers to consume. Satiated by the abundance of food, the warblers cannot control the budworm population, and an outbreak begins.

Baskerville also noted that after World War II, the outbreaks in New Brunswick seemed to be most prevalent in the younger stands, especially fir stands between the ages of 30 and 40 years.[14] Instead of acute outbreaks in isolated mature stands, as before the war, there were chronic but more frequent outbreaks of budworm across the landscape in younger stands. What was the reason for this abrupt shift in outbreak regimes before and after World War II? Why could the warblers no longer control the budworm populations?

Baskerville proposed that World War II marked a significant change in the way timber was harvested that made the forest more vulnerable to budworm outbreaks and reduced the ability of the warbler populations to control them. During World War II, tanks with tracks and chemical insecticides to prevent mosquito transmission of malaria to Marines in the Pacific Islands were developed. After World War II, skidders were developed that used the same tracked technology as the tanks, and the timber industry began to use chemical insecticides. With these new tools, timber harvesting rapidly switched from hand, horse, and oxen logging to a highly industrial machine. As a result, large clearcuts became more widespread in New Brunswick than before the war, during the horse and oxen days. Balsam fir trees, the preferred food of budworms, became established from seeds in these clearcuts. The stand grew into a large monoculture of balsam fir trees of the same age, whereas before the war harvests were smaller and not as likely to be clearcuts. The widespread use of clearcutting synchronized the age class distribution of the forest to younger ages of balsam fir across the landscape. The forest was now perfect for budworms but less so for warblers, which prefer the more complex canopy in older stands.

The industry also began an intense program of spraying insecticide in the hope of controlling budworm completely and preventing isolated but intense outbreaks. The insecticides applied to control budworm weakened warbler populations both by direct poisoning and by reducing the budworm populations that the warblers depended on. Increased deforestation of warbler wintering grounds in the tropics, spurred by the same technological developments as in the north, may also have reduced the return of spring migrants back to their breeding grounds in the North Woods.[15]

Consequently, changes in harvesting practices after World War II simultaneously caused the decline of warblers and the expansion of large areas occupied by single-age cohorts of balsam fir, the preferred food of budworms. Budworm outbreaks became synchronized in these young balsam fir stands, which became the most common forest across the landscape. New Brunswick is now in what appears to be a permanent and chronic spruce budworm outbreak that keeps the forest in younger age classes. The timber industry and the environment have incurred additional costs as spraying programs, begun in the 1950s as chemical substitutes for what warblers do for free, continue unabated. Unfortunately, things are now in a state where it is difficult to get the forest back into a diversity of ages across the landscape that would include the mature stands of spruce that warblers prefer: The budworm outbreaks in younger stands prevent them from growing into mature stands. Indeed, the paucity of warblers themselves and their weakened predation on budworms is one factor that prevents it. Despite the assurances of some industry and public foresters that best forest management is their prime objective, warbler "management" is not high on the agenda anywhere. If the populations of warblers decline elsewhere as they have in New Brunswick, spruce budworm outbreaks may become chronic and widespread in spruce and fir stands across the North Woods.

We have what seems at times to be an unlimited capacity to change the environment, but we are always surprised by the unexpected ways nature responds. When it comes to nature, we seem to lack the ability to learn from our mistakes, and so we are condemned to repeat them. The unexpected directions nature takes often stem from the loss or at least large decline of a group of species we didn't know were important or didn't think would be affected by our actions. MacArthur's warblers are a case in point. As Aldo Leopold said, when tinkering with something, don't throw away any of the parts. It turns out that warblers may be the part we need to help us control budworms.

14.

The Dance of Hare and Lynx at the Top of the Food Web

The large population cycles between lynx and snowshoe hare are iconic symbols of the North Woods and its cousin the Boreal Forest, but their causes remain obscure.

In 1921, Charles Sutherland Elton, then an undergraduate at Oxford, joined the first of three expeditions to study the ecology of Spitsbergen Island, north of Norway. On these expeditions, Elton was a field assistant to Sir Julian Huxley, grandson of Thomas Huxley. Sir Julian was one of a small group of naturalists, ecologists, statisticians, and mathematicians who, in the 1920s and 1930s, unified Darwin's ideas of natural selection with Mendel's laws of genetics.[1] This unification became the modern theory of evolution largely accepted by the scientific community today. While in Spitsbergen and also on a later expedition to Lapland, Elton noticed the wide fluctuations and overland dispersals of populations of lemmings. In his field diary one night, he wrote that he "lay out on a river bank and watched lemmings swim across one by one in the faint darkness of the Northern summer."[2] Much of Elton's later research began with simple natural history observations such as this; in fact, Elton's career goal was to develop the new science of ecology as "scientific natural history."[3]

Elton and others quickly noticed that the population fluctuations of lemmings, voles, and other mammals occur on a regular basis. The series of regular fluctuations became known as a population cycle. Population cycles are characterized by two features: their periods and their amplitudes. The period is the span of time between peaks in the cycle, and the amplitude is the difference between the population size at the peak and the average population size across several fluctuations. Population cycles and the factors that control their periods and amplitudes became the central scientific problem of Elton's scientific life. Before Elton, most biologists, including Darwin, thought that the sizes of populations were largely constant, with minor random variation due to weather and other external factors. In the 1920s, Elton and others began to think that the regularity of cycles—3 or 4 years in some species, 10 to 12 years in others—was due to some underlying but as yet unknown law of the internal dynamics of populations. This was heresy at the time, but it became and still remains a canonical problem of ecological theory.

Elton's magisterial investigation of the population cycles of small northern mammals, *Voles, Mice, and Lemmings*,[4] is in many ways the founding document of population ecology. Every naturalist and ecologist should read this book. Besides writing in an engaging style, Elton showed in this book how theories of population dynamics can be developed from the natural history of these animals and their predators. Elton noted that regular and large population cycles are mostly confined to northern regions and are very uncommon south of the North Woods or the equivalent forests in Europe. Why these cycles are associated with northern environments remains an open problem even today. Elton suggested that additional data on cycles of other northern mammals may shed some light on the problem.

Elton got his wish for more data when one of the leaders of the Spitsbergen expeditions, George Binney, became a biologist with the Hudson's Bay Company and enlisted his expertise as a consultant. From this consultancy, Elton obtained records of the number of lynx trapped, bought by trading posts, and shipped back to England over the past several hundred years. After correcting for differences in accounting procedures from trading post to trading post, Elton and his colleague Mary Nicholson demonstrated that these records were reasonably good proxies for population densities. From these trapping and shipping records, Elton and Nicholson compiled and published the first time series of lynx population cycles.[5]

Because lynx prey almost exclusively on snowshoe hare, Elton soon realized that to understand the cycles of lynx populations he would have to also understand the population dynamics of the snowshoe hare (also called the varying hare, or snowshoe rabbit). He began to study data compiled by D. A. MacLulich and others on hare and lynx populations.[6] Along with his colleagues Dennis and Helen Chitty, Elton also began the Canadian Snowshoe Rabbit Enquiry, a long-term and continent-wide study of population cycles of snowshoe hare across the boreal forests of Canada and the North Woods of the Great Lakes Drainage Basin.[7] Their graphs of the population dynamics of lynx and snowshoe hare, which can be found in almost every introductory textbook on ecology, have become iconic symbols of the boreal forest and the North Woods.

Elton and the Chittys noticed that these population cycles of lynx and hare had several peculiar and asymmetric features. First, the populations fluctuated fairly regularly over two orders of magnitude on a 9- to 11-year cycle in boreal regions, but the magnitude of these cycles differed from region to region. For example, the number of lynx trapped from the Upper Saskatchewan River was only 150 in 1833 but skyrocketed two orders of magnitude to 15,975 by 1838. In contrast, lynx and hare populations in the more southerly North Woods of the Great Lakes region only doubled or tripled from lows to peaks. Second, cycles were not a nice symmetrical sine wave in which the low phase of the popu-

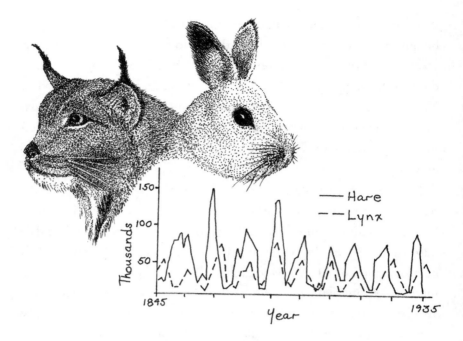

lation cycle was a mirror image of the high phase. Instead, there was a decided asymmetry between population peaks and lows: The high point in both cycles was usually a sharp peak lasting for a single year, followed by steep crashes to population lows that lasted for 3 or 4 years. Third, fluctuations of the lynx population were delayed behind those of the snowshoe hare by several years. It is as though the hare were leading the lynx through a dance of some sort, but one for which we don't yet know the tune. Finally, despite these asymmetries in the cycles through time, the population fluctuations of both lynx and hare were synchronous over enormous regions, such as the entire Winnipeg Drainage Basin east through James Bay and south through the North Woods of the Great Lakes Drainage Basin.

Most research until recently has tried to infer the causes of these cycles by mathematical analysis of time series of populations and their correspondence with environmental factors.[8] In a joint mathematical

analysis of both lynx and hare cycles,[9] Nils Stenseth and colleagues concluded that the lynx populations are regulated almost entirely by the
supply of hares. The growth of the lynx population is in turn a major
factor controlling the amplitude of the hare population cycle: The faster
the growth rate of the lynx population, the smaller the amplitude of the
hare cycle.[10] Hare population cycles are also controlled by other predators such as great horned owls, goshawks and other raptors, red fox,
and wolves, but less is known about how these other predators relate to
the hare cycle. But unlike lynx, hare population cycles also seem to be
driven by the abundances of its plant foods, such as birch, willow, aspen,
red osier dogwood, and other species. These forage species are partly
depleted during the peak of the hare's cycle but recover later, so the hare
cycle involves three trophic levels (plants, hare, and predators including
lynx), unlike the lynx population cycle, which involves only two trophic
levels, the predator and its prey. The coupled hare–lynx cycle does not
appear to be a simple symmetric predator–prey cycle but something
decidedly asymmetric.

It seems that further attacks on the problem should focus on the
hare cycle, because it was pretty clear that whatever explained the hare
cycle would explain the cycle of the lynx that depended on them. An
experiment that manipulated food and predator density, each alone and
then together, would help sort out the three-trophic-level hypothesis
of Stenseth and colleagues, but this experiment would have to be on a
very large scale commensurate with the home ranges of hares and the
landscape they lived in. Charles Krebs and his colleagues planned and
implemented such an experiment in the wild country of the Kluane
Basin in southwestern Yukon, near the knuckle of the Alaskan panhan-
dle.[11] Here, they constructed a square-kilometer exclosure with electri-
fied fences to exclude mammalian predators of the hare, but the bottom opening of the fence was large enough to let hare through. The
fence could not exclude birds of prey, so it did not prevent predation

entirely, but the fence decreased predation substantially nonetheless. To test whether the abundance of food caused the hare population to cycle, they added rabbit chow to another square kilometer of forest. And to test whether food and predators had an additive effect on hare abundance and cycles, another exclosure was constructed to exclude predators, but inside it they also added rabbit chow. Because rabbit chow is a decidedly unnatural food, an additional square-kilometer block was fertilized with nitrogen, phosphorus, and potassium (the three major nutrients plants take up from the soil) to boost productivity of the natural plant foods of hare.

If predators and food supply caused the cycles, then predator exclusion and food enhancement should stop the cycles, or at least dampen them. Three blocks, each 1 square kilometer, were also marked to serve as baseline controls from which population deviations inside the exclosure and food blocks could be assessed. It was simply not practical to replicate the large exclosures, so Krebs and colleagues hoped for very large differences between hare populations inside them and the controls, differences that could not reasonably be attributed to any other factors.

Their hopes were granted. Excluding mammalian predators doubled hare population density above the control blocks, mainly by greatly reducing mortality. However, adding food tripled it throughout the cycle. If adding rabbit chow and excluding predators acted independently of one another, then we would expect the block with no mammalian predators and added food to increase hare density five- to sixfold, but instead hare density increased by eleven times in this block. Adding rabbit chow somehow multiplied the effect of excluding predators (more on this in a moment). Although fertilization increased plant growth and supplies of forage to the hares, the hare population density in the fertilized block increased only slightly above the control.

These treatments increased the density of hare populations and even sometimes the amplitudes of the cycles. But if either the scarcity of food or predation by lynx caused the cycles, then augmenting food supplies with rabbit chow and excluding lynx would have stopped the cycles, but they didn't. Food and predators appear to control the amplitude and timing of the cycle, but the existence of the cycle might lie in the number of hares born each year and stresses within the hare population itself.

What mechanisms affect the hare's reproduction and body condition, and how do these mechanisms relate to the effects of predators and food supply? One possible stress is the increased social disruption and hormonal changes in females during the population peak, which could decrease reproductive output.[12] In addition, even though lynx are successful in killing hares only 40 percent of the time, the additional 60 percent of failed attempts by lynx may stress the hares almost continuously, also causing changes in hormonal regulation of reproductive cycles in females. At the beginning of the growth phase of the hare population, food is abundant, and the lynx population has yet to increase. Life is good, stress is low, and snowshoe hares can produce four litters per year with four leverets (as the young hares are called) per litter. But as the hare population grows, social stresses begin to build, and the lynx population also begins to increase. Even before the peak of the population cycle, stresses to the female hares result in hormonal changes that curtail first the fourth and then the third litter. The weight and condition of the young from the first two litters also decrease, leading to lower survival rates. But these low reproductive rates last well into the decline phase, and it may take several years for the females to recover normal hormonal levels or for a sufficient number of young, unstressed females with normal hormonal levels to dominate the population. The prolonged hormonal changes in previously stressed reproductive females may be one reason why the population low lasts for several years. Indeed, the population does not begin to recover until the stress levels of the females drop to the point where they again can begin to produce a third or fourth litter.

Delays in the recovery of the food supplies from heavy browsing by hares might also keep the population in an extended low phase and thus also lead to cycles. When stems of aspens and birches are browsed by hares in winter, their roots and stumps sprout abundant juvenile shoots the next summer. One winter a number of years ago, we had a particularly large snowshoe hare population in northern Minnesota along with a deep snow cover with a crust that was thick enough to support snowshoe hares but not lynx or wolves, which foundered in the deep snow when they broke through the crust. The hares debarked thousands of aspen and birch stems at about the height of the snow crust above the ground that supported them. In spring, these mature, debarked stems died, but they were replaced by an even greater number of new juvenile shoots. However, the hares may not have considered these to be suitable food, because the next winter they were not as heavily browsed as the twigs on the older aspen and birch saplings.

Juvenile birch and aspen produce resin glands along their stems that contain distasteful and toxic substances such as **terpenes** and phenolics; once the stem gets above the usual browse height of hares and enters a "safe zone," so to speak, it stops producing these glands and instead allocates the energy to produce lateral branches.[13] John Bryant of the University of Alaska extracted the phenolics and resins from juvenile shoots of paper birch, quaking aspen, balsam poplar, and green alder. He painted these extracts on shoots of feltleaf willow, another preferred food of snowshoe hares that does not produce these resins, and allowed snowshoe hare to browse them along with untreated control twigs ad libitum in a sort of cafeteria experiment. The proportion of twigs that were browsed declined exponentially with increasing concentrations of the resin painted on them, clear evidence of the deterrence power of these compounds.

Bryant also told me that he has seen hares snip off the current year's growth of birch twigs where the resin glands are most dense and eat the older portion below it, which contained fewer glands. The clipped twigs litter the snow surface alongside copious hare tracks. I've not seen this foraging behavior in northern Minnesota. Perhaps our hares are not as clever as Alaskan hares, or perhaps there is some geographic variation in birch chemistry or hare behavior between here and Alaska. In any case, Bryant proposed that a delay in the recovery not of the quantity but of the quality of food may also keep hare populations and their reproductive output at low levels for several years, thus leading to a population cycle.

Despite these changes in food abundance and quality during a snowshoe hare cycle and despite the increased population density in the plots with augmented food supplies in the Kluane experiment, hares at Kluane rarely died of starvation. Instead, poor quality or sparse food at the onset of the decline in hare populations may make hares more susceptible to capture by lynx and other predators. Hares, like moose and other northern herbivores, prefer to browse where food patches are adjacent to conifer cover because in such places they have a safe haven when a predator approaches. But when such food patches become overbrowsed or dominated by juvenile stems with toxic substances, as in the Kluane control plots, the hares may venture farther from cover to obtain sufficient food to meet daily energy needs. This not only increases the probability that they will be found by predators but also causes them to expend more energy, causing their condition to deteriorate so that escape from predators becomes less likely.[14] Whether hares extend their foraging distances from cover when juvenile stems with phenolics and terpenes dominate their food supply could be easily tested today by outfitting hares with new miniature GPS collars. This interaction between the distribution of food and exposure to predators may explain the large response of the hare population in the Kluane experiment where food was added and predators were excluded, a response larger than would be expected by combining the independent responses of both together.

The Kluane experiment is one of the largest experimental manipulations of herbivores and their food supply. It required heroic efforts to maintain this experiment: The electric fences had to be checked every day during winter, often in temperatures as low as −45°C. The experiment provided much valuable insight into the factors that control hare populations in a single valley. But hare and lynx population cycles are also synchronized over vast regions such as the entire region west of Hudson's Bay. No possible experiment could be done to test hypotheses of why the cycles are synchronized over such large areas; it would be impossible to fence off the entire province of Manitoba, for example. Tests of hypothesized mechanisms that might synchronize population cycles over such vast areas can be done only by examining correlations between the cycles and mechanisms that work on very large scales. Correlations between two things do not necessarily mean that changes in one thing cause changes in the other; only experimental manipulation of hypothesized factors, such as excluding lynx from a hare population as in the Kluane experiment, can explicitly test whether one thing causes another. But when we can't physically do such an experiment, correlations combined with sound reasoning are the best we can do. This is the case for the problem of what factors synchronize hare and lynx population cycles over very large regions.

One such mechanism could be periodic climate fluctuations, which in turn could cause food supplies for hares to fluctuate. During peaks in hare populations, their preferred deciduous foods become scarce as they browse them. The hare may then begin to browse the main shoots of white spruce seedlings and saplings. Tony Sinclair and colleagues demonstrated that when the main shoot of a white spruce seedling or sapling is browsed by hare, an unbrowsed side shoot will turn upward in subsequent years and replace the original main shoot.[15] Simultaneously, diameter growth of the sapling decreases until a new main shoot is established. During this period of suppressed growth, the rings at the base of the tree will become extremely close together and look like a single, very dark ring. Sinclair and colleagues therefore could use tree ring analyses of cross-sections of spruce trees to extend the snowshoe hare population record back to the early 1700s. They then found that periods of dark rings and high snowshoe hare populations were correlated with periods of high sunspot activity. When sunspot activity is high, the output of heat from the sun is also high, and the climate tends to be warmer. It is reasonable that plant growth, and hence food production for the hares, would be greater during periods of high sunspot activity. Sinclair and colleagues point out that the period of the solar sunspot cycle is nearly the same as the 10- to 11-year period of the hare–lynx cycle and that the correlation between solar sunspot cycles, climate, and tree ring growth has been well documented elsewhere. The solar sunspot cycle may therefore entrain hare cycles over large regions through the effects of climate on tree growth, especially the growth of juvenile white spruce during peaks in the hare populations.

To test their hypothesis further, they suggested extending the tree ring–snowshoe hare record back through the 1600s, 1500s, and 1400s. During this time, there were two prolonged periods of low sunspot activity, one between 1400 and 1510 and then another between 1645 and 1715. The climate was cooler during these times of low solar heat output. If sunspot cycles control the growth of both white spruce and hare populations through their effect on regional climate, then during these periods we should not see any of the dark rings of spruce trees produced during peaks in the hare populations: The hare population cycles could have been shut down by the lack of food during these cool periods. It is a rare spruce tree that would live that long, especially in fire-prone areas such as the western Great Lakes region (see the essays in Part V). However, there may be isolated populations of very old spruce that may contain the data within their rings to confirm or refute this hypothesis near the treeline in northern Quebec, Ontario, Manitoba, and Saskatchewan.

Since Elton's pioneering studies of the natural history of population cycles, we have learned much about their causes, but open questions and loose ends remain: How far do hare travel from cover when food supplies diminish? Does their survival decrease away from cover? Is the hare cycle dampened where there is a greater diversity of predators? How do snow conditions modify the hare–lynx cycle? Will regional tree growth and the hare cycle continue to be entrained by sunspots even in the coming warmer climate? We also need additional large-scale experiments such as the Kluane experiment but now with controlled clipping, as Kjell Danell and I did in smaller exclosures in Sweden (see Essay 11, "What Should a Clever Moose Eat?") to test whether Bryant's hypothesis applies to entire hare populations and at larger landscape scales. To answer all these questions, we will need new natural history observations of the life cycles of lynx, hare, and the hare's preferred food species. Elton was always guided by natural history observations, and his approach remains as valid today as it was when he was on Spitsbergen nearly a century ago. Elton once said, "Unfortunately, nature cannot be understood by pretending that it is simple."[16] Perhaps he should have said "Fortunately" instead of "Unfortunately," because it is the complexity of nature as it is revealed through observations of natural history that forces us to take our theories and hypotheses lightly.

PART IV
Pollinators, Flowers, Fruits, and Seeds

Summer is the time for plants in the North Woods to produce seeds that pass along their genes to the next generation. The growing season between frosts is only from June to mid-September in most of the North Woods. This is a short time to produce seeds, so it is imperative for the plants to attract pollinators quickly once flowers open. Pollination is the combination of genes contained in the seeds, produced by the ovaries in the female part of the flower, with genes carried by pollen, produced in the male part of the flower, known as the stamens. Other than mosses, liverworts, ferns, and a few others, all plants including all trees and all understory wildflowers need pollination. Whereas some plants can fertilize themselves with their own pollen, others need to be cross-pollinated with pollen from other individuals in the same species. Cross-pollination is necessary when male and female flowers are produced separately on different plants or when the pollen in the male parts of one plant ripens at a different time than the seeds become ready to be pollinated on the same plant.

In some species, pollen is transferred from one plant to another by the wind, but in many others pollen must be transferred from one plant to another by insects. These are the plants that produce showy flowers, rich smells (not all of them pleasant, as we shall see), and nectar to attract insects. Insects have also evolved traits to transfer pollen, such as the hairs on bees, to which the pollen sticks while the bees fly from one flower to another. When insects cross-pollinate plants, they sustain the population of plants from which they receive nectar and on which their eggs are laid and larva feed, so both plants and pollinating insects benefit one another. These mutual adaptations are classic examples of coevolution. But the seeds must also be dispersed to suitable sites to germinate and produce a new generation of seedlings. Fruits are produced to attract many birds, squirrels, and bears, which consume the fruits but defecate the seeds elsewhere unharmed, another example of mutually beneficial coevolution. On the other hand, animals such as crossbills and squirrels can also destroy the seeds if they crunch them with their beaks or teeth. Conifers have cones armored with spikes to defend the seeds against seed predators. But predators that can overcome these defenses are at a competitive advantage and pass down their traits to their descendants. This is an example of a coevolutionary arms race, which we have also seen in Essay 12, "Tent Caterpillars, Aspens, and the Regulation of Food Webs." Because of the short northern growing season, selection pressure for coevolution between flowers, pollinators, seed dispersers, and seed predators is strong in the North Woods and has produced a wide variety of adaptations between plants on one hand and insects, birds, and mammals on the other.

Flowers, fruits, cones, and seeds are expensive investments of energy and nutrients, which are usually in short supply because of the short growing season and the nutrient-poor soils common throughout the North Woods. Energy and nutrients are also needed by leaves to gather more energy, by stems to get the leaves up to the light, and by roots to gather more nutrients. There are therefore serious tradeoffs between supplying energy and nutrients to seeds, fruits, and cones instead of to leaves, stems, and roots.

How should plants allocate their resources to reproductive and vegetative structures? When and for how long should flowers bloom and fruits ripen? Why does one species defend itself against seed predators while another allocates large amounts of precious sugars to fruits to attract seed dispersers? These are some of the questions raised by the natural histories of flowers, insects, birds, and mammals, which we will consider in Part IV.

15.

Skunk Cabbages, Blowflies, and the Smells of Spring

How the smell of a rotting carcass attracts carrion flies to skunk cabbage and fools them into pollinating the flowers.

After a long, cold, and white winter with the sharp, clean smell of arctic air, many of us are ready for some brilliant greens and the sweet smells of cherry blossoms and wild roses and the earthy smells of wet soils as the melting snows percolate into the ground. One of the first plants to emerge from the snows and show its brilliant green leaves is the skunk cabbage. Although the foul smell of the skunk cabbage is not what most people have in mind when they think of spring, it is a key adaptation in an interesting network of relationships with other members of the ecosystem in which the skunk cabbage lives.

The skunk cabbage can be seen in abundance in forested valley bottoms throughout the northern hardwoods of eastern North America and even into the boreal regions of Ontario and Quebec. Its large, brilliant green leaves give it a somewhat tropical appearance, and for good reason. The family of which it is a member—the Araceae, or Arum family—is largely tropical, with only about a dozen species in North America and only four in the North Woods. Skunk cabbage, along with three close relatives, Jack-in-the-pulpit, sweet flag, and wild calla lily, are the northernmost representatives of the Araceae. The latter three species range as far north as the border with Ontario, whereas the skunk cabbage's range extends farther northward.

The flower of the skunk cabbage, like that of the Jack-in-the-pulpit, is not showy. It's a large purple or brown hood, or spathe, that encloses a clublike stem, called a spadix, that holds the male stamens, which produce pollen, and the female pistils, which produce the seeds. We normally think of flowers as being showy and sweet smelling to attract their pollinators, not drab and foul smelling like that of the skunk cabbage.

Bees come to mind most often when we think of pollinators, but bees have not yet emerged from their hives when the skunk cabbage begins to flower as the snow melts around it. The skunk cabbage has to make do with whatever insects are around early in spring. These early spring insects make their living by consuming and laying their eggs in the rotting flesh of animal carcasses, which are exposed as the snow melts. Among these are carrion beetles and the common blue bottle fly or common blowfly. The blowfly's larvae survive the winter in the crevices of the dark bark of oaks and maples. The dark bark heats up with the warmer and higher early spring sun as the snow melts, causing adult blowflies to develop from the overwintering larval stage. The adults quickly take flight and seek a rotten carcass for their first meal and a place to lay their eggs.

This is where the foul smell of the skunk cabbage (hence its name) enters the story. Skunk cabbage produces various aromatic compounds that mimic a mixture of skunk, garlic, and rotting flesh. While following cues of foul smells from dead mammals wafting in the spring breezes, a blowfly is occasionally diverted into the skunk cabbage's hooded spathe from which some of these same smells emerge. As the blowfly rummages around in the purple hood around the flower, it covers itself with pollen. Eventually it gives up and leaves the shelter of the spathe when it realizes that this is not a nice, juicy dead deer. Insect brains are not known for their complexity, being only a few nerve cells, and the blowfly does not seem to learn too quickly that this is not a source of food. Consequently, as the blowfly emerges from the shelter of the spathe it may pick up the scent of another nearby plant, especially because skunk cabbages generally grow in large groups. The blowfly will then be fooled into entering and pollinating a nearby skunk cabbage with the pollen from the previous plant.

Many plants produce rewards to attract potential pollinators or other insects that benefit them. For example, some flowers produce nectar as a reward for insects that pollinate them. But skunk cabbages seem to attract insect pollinators by deceiving them into thinking they are getting a meal of juicy rotten meat in which they can also lay their eggs. There doesn't seem to be any reward for the poor blowfly that enters the spathe and deposits pollen; the reward goes entirely to the skunk cabbage, which gets pollinated.

Or is the pollinating insect rewarded by something else the plant provides? Spring is a cold time of year, especially when skunk cabbages bloom, while there is still snow on the ground. All insects are cold blooded. This does not mean that their blood is always cold or that their body temperature fluctuates entirely at the whims of the environment but that they must regulate their body temperature by their behavior.[1] The most obvious behavior is to seek warm environments. For example, butterflies on cold mornings spread their wings toward the sun to gather heat; the wings become solar collectors that warm the blood in the wing's veins and transfer the heat to the butterfly's body. Skunk cabbages may literally be hot spots in the landscape, attractive to insects because they are one of the few plant species that actually generates heat by respiration, much as warm-blooded animals do. This member of a tropical family actually brings tropical warmth to the late winters and cold springs of the North Woods.

Roger Knutson, a botanist at Luther College in Decorah, Iowa, and Roger Seymour of the University of Adelaide in Australia have each found that the air temperature in a skunk cabbage spathe remained nearly constant and warmer than ambient air temperature: The air in the spathe generally remained in the 10-degree range between 15°C and 25°C, even when ambient air temperatures fluctuated throughout a 30-degree range, from as low as −15°C to +15°C.[2] The temperature of skunk cabbage flowers is warm enough to melt the snow covering it and surrounding it.

In an elegant experiment, Knutson showed that this was not warm air trapped in the hooded spathe but air heated by the elevated respiration rate of starch reserves in the huge roots of the plants. Knutson showed this simply by clipping the spathe from the starchy root. The temperature of the spathe immediately plummeted once the connection with the starch was severed. Knutson also measured oxygen consumption and Seymour measured carbon dioxide production by several spathes maintaining a temperature of 20°C (approximately room temperature) at an air temperature of 0°C. The temperature of the spathes, their oxygen consumption (which drives the heat-producing respiration reactions), and their carbon dioxide production (which results from the burning of starch reserves) were equivalent to those of a small mammal, such as a red-backed vole. Oxygen consumption and carbon dioxide production increased as the air temperature got colder, meaning that the plant was able to sense the cold environment and burn starch faster to maintain the high temperature of its spathe, exactly as warm-blooded animals do.

This warmth creates small convection cells in the air above the plant, which disperse the skunk odor, but the heat produced by the skunk cabbage may also be a reward for blowflies who enter the spathe. The insects are quickly warmed and take some of this warmth with them when they reenter the cold world outside the spathe. Even though the insect is deceived by the skunk cabbage smell (and wastes time and energy that would otherwise have been spent on finding a rewarding meal of rotten meat), the cost of this deception may be entirely offset because it is also warmed and does not have to spend its own sugar reserves warming itself.

It is almost axiomatic of ecology that once two organisms begin inter-acting, others soon learn how to exploit that interaction. Knutson and others have found spiders lying in wait at the opening of the spathe to snare whatever blowfly happens to come along. It is possible that these spiders and even other insects mate and hatch eggs in the warm and moist environment of the hooded flower of the skunk cabbage.

The skunk cabbage is a small food web in and of itself, consisting of at least a plant, the blowflies and other spring flies, and the predatory spiders, among other possible predators. Three trophic levels in this tiny food web! By studying one trait of the skunk cabbage—or any species, for that matter—we uncover a surprising and almost never-ending web of connections to other forms of life.

Nevertheless, many questions remain unanswered: What are the photosynthetic rates of skunk cabbage? Are these rates very high, in order to replenish the huge reserves of starch needed to heat the air? Is photosynthesis in skunk cabbage limited by nutrients or light or both? How many plants does a blowfly visit and therefore pollinate before it learns that this is not a source of rotting flesh? Are skunk cabbages growing in the vicinity of an animal carcass more frequently visited by blowflies or less frequently visited? Each bit of knowledge we gain by answering one question opens up new questions and paradoxes. The interaction between the skunk cabbage and the blowfly may seem an unlikely entry into the scientific study of the natural world, but it is a fascinating one nonetheless.

16.

When Should Flowers Bloom and Fruits Ripen?

Coevolution of flowering time, pollinators, fruit ripening, and seed dispersers of juneberries.

In the North Woods of Minnesota, we begin picking juneberries in the middle of July. In slightly warmer climates the berries ripen in June, hence the name, but in colder climates the snow-white flowers of juneberries make a splendid display in late May or early June. The best way to find the ripe berries in July is to make a mental note of where you saw small trees with clouds of white flowers 6 or 7 weeks ago. Some of these trees may be wild plums, but there's a very good chance they will be juneberries. The leaves of juneberry emerge copper orange as the flowers begin to bloom but gradually turn green as the bloom subsides. The spring of 2015 was a banner year for juneberry blooms, the best I've seen in 30 years. Almost everywhere I looked, there was a mosaic of white flowers and coppery leaves against a background of pastel greens emerging from birch and aspen. This spectacular bloom made me realize that there are more juneberries in the forests around here than I had previously thought. Or perhaps the abundance of juneberries has been increasing in the past several decades without me realizing it.

Along the Atlantic Seaboard this beautiful small tree is called shadbush because there the trees bloom when the shad make their mating runs from the ocean up into the rivers. In the Great Plains, they are called by their Indian name *Saskatoon* (from which the town in Saskatchewan gets its name). Indians and voyageurs pounded saskatoon berries with bison or moose meat and fat and then stuffed them into the intestines and stomachs of these animals to make pemmican. Pemmican was the "power bar" of the voyageurs, fueling the grueling sixty strokes per minute they used to paddle their canoes upstream hour after hour, in rain and sunshine. *Serviceberry* is also a common name for juneberries, especially in the western Great Lakes region and Canada. This name appears to be an Anglo-Saxon corruption of the Latin *Sorbus*, which is the genus of an entirely different tree, the mountain ash, that looks nothing like the juneberry except for also producing reddish berries.

The juneberries comprise about ten species that belong to the genus *Amelanchier*. The distribution of most juneberry species is centered largely on the Great Lakes, except for the saskatoon, which is found in the prairies of western Minnesota through the northern Great Plains, and the downy serviceberry, which is found throughout eastern North America from the north shores of the Great Lakes to the Gulf of Mexico. The ranges of the remaining species are found in the North Woods from northeastern Minnesota to Maine.

Plant taxonomists disagree on the number of species in the genus *Amelanchier*, which leads to some unresolved confusion about the structure and evolution of this genus. Carl Rosendahl, a professor in the 1920s in what used to be the University of Minnesota's Department of Botany, described eleven species in Minnesota,[1] but Welby Smith of the State of Minnesota Department of Natural Resources today recognizes eight species.[2] Smith lumps four of Rosendahl's species into seven remaining species but then recognizes one additional species. You are probably now as confused as the plant taxonomists. The reason for the difficulty in identifying distinct species is undoubtedly the strong propensity for the plants to cross-pollinate and hybridize with each other. Evolutionary biologists sometimes call such an assemblage of species a **hybrid swarm**. Hybrid swarms may represent a very early stage in the evolution of new species, as the populations are only beginning to be genetically and reproductively distinct from one another.

Hybrid swarms are common in the rose family, of which juneberries are members, along with hawthorns, raspberries and blackberries, apples, cherries and plums, the aforementioned mountain ash, and several others. Hybridization is extensive within and even between these different genera, which is why horticulturists can produce so many different varieties of these fruits. Michael Pollan documents the fascinating story of how many of the modern apples developed after a few varieties of English and European apples cross-pollinated with America's wild crabapples, which are in the same genus, as the pioneers moved west of the Alleghenies and planted orchards from Ohio to Minnesota.[3] This cross-pollination of apples with wild crabapples resulted in a huge evolutionary diversification of the apples in North America, resulting in the enormous variety of heirloom varieties available at grocery stores and especially farmers' markets.

This extensive ability to hybridize implies that all the flowers of these species must have some traits in common, the most obvious being what botanists call inferior fruits. This does not mean that botanists dislike eating these fruits, because they are decidedly superior in taste. Instead, *inferior* refers to the fact that the ovaries and the fruits that develop from them form below rather than above the petals and sepals of the flowers. The rough structures you see at the end of apples, pears, peaches, cherries, and juneberries on the side opposite the stem are the remains of the sepals of the flower. The fruit is the swollen ovary that was pollinated by bees and other insects earlier in spring.

The snow-white flowers of juneberries are arranged along a short, flowering shoot. Each flower has five elegant and slender petals that open nearly synchronously throughout the tree.

Trees that synchronize their flowering times flood the air with their aroma all at once and therefore maximize their chances of attracting bees and other pollinators for cross-pollination. If the initiation of flowering stretches out over several weeks, only a few flowers will be open at any one time, and bees may ignore them.

Plants whose flowers open synchronously stand a better chance of being pollinated and therefore produce more berries containing the seeds to produce descendants than asynchronously flowering plants, which may not get pollinated at all. The descendants of these pollinated plants inherit their parents' genes controlling the synchronization of flowering; these genes then spread through successive generations. But if different populations flower at different times, they minimize cross-pollination and eventually diverge into different species, a process known as **speciation**. This is an example of how Darwin's theory of natural selection increases species diversity. The reason juneberries form hybrid swarms is that all of the different species flower at roughly the same time, giving bees the opportunity to transfer pollen (and hence genes) from one species to another and thereby inhibiting speciation to some degree. Natural selection has not completely separated flowering time between the different species of juneberries, resulting in the extensive hybrid swarm of species in the *Amelanchier* genus.

Even though flowers emerge nearly synchronously on a given juneberry tree, the fruits ripen asynchronously. Ripening of the fruits stretches out over several weeks from the middle of July through the middle of August. When we get enough summer rains, the dark, blue-black berries are large and juicy. They taste like a mix of blueberry and cherry and also make great pies and jam.

The purpose of spending so much energy and nutrients producing juicy, edible fruits is to entice birds and other animals to eat them and disperse the seeds elsewhere. Delayed and asynchronous ripening of the berries through July and into August increases the chances of dispersal of the seeds by birds for several reasons. At any one time from mid-July to early August any tree will have green, red, and blue-black berries in various stages of ripening. The multicolored berries on a single tree are visually striking and undoubtedly attract the attention of hungry birds, as they also attract my own attention.[4] The early ripening berries contain up to ten viable seeds, but the late-ripening berries contain only one or two viable seeds.[5] This maximizes the chance of seed dispersal: Birds will be attracted to the trees at the early stages of fruit ripening because of the striking display of multicolored fruits in various stages of ripening, and these fruits are the ones that produce many seeds to be dispersed.

In late August and early September, the number of ripe berries on a tree declines, and those that remain are all blue-black, so the tree is not as visually striking as earlier, when fruits were green, red, and blue-black. Therefore, the birds switch their attention to later-ripening species such as mountain ash. The late-ripening berries also contain fewer seeds. So there is a definite advantage to having the fruits with the most seeds ripen early, when the birds are most likely to turn their attention to the trees with fruits of many colors. The remaining fruits that ripen later are a sort of "bet hedge" in case the earlier fruits did not attract birds, but it is best to invest only a few seeds in these fruits because their chance of being eaten is less.

Darwinian natural selection of juneberries therefore synchronizes blooming to attract bees and other insects that pollinate flowers, but it also independently desynchronizes ripening to attract the birds that disperse the seeds.[6] Together, the striking asymmetry of synchronized flowering and desynchronized fruit ripening maximizes the number of juneberry descendants. The strategy that maximizes reproduction at one stage of a plant's life cycle may be the opposite of what works at another stage in the life cycle. There is no "miracle" strategy that always works best, no one-size-fits-all solution to the problem of maximizing the number of descendants that carry the parent's genes. In natural selection, anything goes as long as it leads to more descendants and more copies of the parent's genes in future generations.

The different evolutionary selection pressures at different stages of an organism's life cycle also place natural selection pressures on other species the organism interacts with, such as bees and birds in the case of juneberries. To obtain pollen, the bee must focus its search image on juneberry flowers when they bloom (they are also helped by the flowers' aroma flooding the air). To obtain juneberries for energy, birds must find the juneberries (they are helped by the multicolored berry crops) and eat the berries that are ripe. The abilities of bees and birds to find juneberry flowers and fruits are in turn partly under the control of *their* genes, so selection for synchronous flowering and asynchronous fruiting in juneberries leads to selection for bees and birds with certain foraging behaviors. Selection for certain traits in a species such as juneberries is ramified throughout the food web to the other species with which it interacts.

Natural selection and evolution bind together all species in the web of life on Earth, just as they bind together juneberries, bees, birds, and berry pickers like me. The web of life is not just the flow of energy and nutrients through the food web. It is the set of morphological and behavioral adaptations brought together by natural selection that determines how species interact, one with the others.

17.
Everybody's Favorite Berries

Evolutionary tradeoffs between investment in leaves and fruits in deciduous blueberries and evergreen lingonberries.

When is summer better than when one stumbles upon a patch of ripe, sweet, juicy blueberries on a rocky ridge with a clear, blue sky overhead? Summer in the North is at its peak in late July and early August, when the blueberries are ripe. Most birds have fledged at least one cohort of young, and sandhill cranes, crows, and blue-winged teal are starting to form flocks for the flight south. A few leaves are beginning to turn red along their margins here and there in the forest and on the blueberry plants themselves. The beginning of autumn and even perhaps a frost are mere weeks away.

Blueberries are in the genus *Vaccinium*, along with cranberries and lingonberries. The most common species of blueberries in northern Minnesota are *Vaccinium angustifolium* and *V. myrtilloides*. *Vaccinium angustifolium* has smooth leaves and is more common in the drier areas than *V. myrtilloides*, which has rather fuzzy leaves and is the more common of the two in bogs. Both are equally delicious.

The *Vaccinium* species are members of the family Ericaceae, which also includes heather, wintergreen, and bearberry, otherwise known as

kinnikinnick. The Indians used bearberry's leaves like tobacco in pipe ceremonies, as described by David Thompson in his journals of his travels across Canada, Minnesota, and Oregon in the early 1800s (see Essay 4).

Blueberries are true berries, having a sweet fleshy fruit that encloses numerous seeds. Huckleberries, often mistaken for and sometimes growing with blueberries, are in a different genus than blueberries but are still a related member of the Ericaceae. Huckleberries are not true berries because their seeds are enclosed in a hard, stony pit. Technically, the huckleberry's fruit is known as a drupe, any fruit with a single pit or hard stone. The actual seed is enclosed inside the pit or stone. Peaches and cherries are also drupes.

Vaccinium is a circumpolar genus, with several species being found in northern regions around the world, including Scandinavia and Siberia. One is the Scandinavian blueberry (*blåbär* in Swedish), sometimes called bilberry, which is found in the understory of pine forests throughout Scandinavia and Siberia. The taste of the bilberry fruit is almost identical to that of our blueberry. Another related species, the lingonberry, is most common in Sweden and in the Maritime Provinces of Canada, where it is called the partridgeberry. Lingonberry can occasionally be found in northern Minnesota, and though not rare here it is also not as common as in Scandinavia. Lingonberries are sometimes called cowberry (*vacca* is Latin for "cow," from which the genus they share with blueberries and cranberries takes its name). Lingonberries are usually made into a slightly tart jam or sauce that is often served in Sweden with a moose roast, much as we serve cranberries with turkey. Lingonberry is an unusual broad-leaved plant in that its leaves are evergreen, whereas blueberry and bilberry leaves are deciduous.

It takes a lot of carbohydrates to make a sweet berry. These carbohydrates must be supplied by photosynthesis in the leaves. Being able to make the amount of carbohydrate the berries and the rest of the plant both need while growing on infertile soils poses several problems for both blueberries and lingonberries. First, the photosynthetic machinery in the leaves uses large amounts of nitrogen and phosphorus, which must be taken up from the soil. However, nitrogen and phosphorus are in short supply in the soils because they must be released from decomposing leaf and twig litter by microbes before plants can take them up; the short decomposing season of northern forests combined with the tough nature of blueberry and especially lingonberry leaves makes them difficult to decompose. Second, the high nutrient and carbohydrate cost of producing berries must be borne at the expense of producing shoots, roots, and even leaves themselves; there is only so much nitrogen, phosphorus, and carbohydrate to go around. Allocating these resources to producing one type of tissue may have to come at the expense of producing another.

One way to get around this allocation problem is to boost photosynthetic rates as high as possible. Deciduous leaves have higher photosynthetic rates than evergreen leaves, and so the problem of supplying carbohydrates may not be as great in blueberries as in lingonberries. On the other hand, because they have higher photosynthetic rates, blueberry leaves need higher protein contents to support the photosynthetic machinery. Their higher protein content makes the deciduous leaves of blueberries more nutritious, and therefore they are eaten more often by herbivores such as mice, moose, or reindeer, compared with the tougher evergreen leaves of lingonberry. The evergreen leaves of lingonberry have their own cost, mainly the diversion of much of the carbohydrates initially produced by photosynthesis to the production of lignin in the cell walls. Lignin toughens the cell walls of the leaves, enabling them to withstand winter snow and ice and summer drought and therefore live several years. Lignin deters browsing by moose because the tough leaves are harder to chew and digest, but it also slows the decay of the leaves when they drop to the soil, and so the release of nitrogen and phosphorus for plant growth is also reduced.

Blueberries therefore have a higher rate of photosynthesis to provide carbohydrates needed for their fruit, but the cost of this high carbohydrate production is the potential loss of the photosynthetic machinery to a hungry moose. On the other hand, the tough, lignin-rich leaves of lingonberry enable it to survive several years of winter snow and ice and summer drought and also deter moose, but this means that less carbohydrate is available for the berries. (This gives lingonberries, as well as cranberries, another evergreen relative, their tart taste compared with the sweet taste of blueberries. On a sabbatical in Sweden, I often made pies with three cups of blueberries and one cup of lingonberries, all of which we picked in the forests behind our apartment. The combination of the two gave a wonderful sweet–tart combination of flavors.)

These plants therefore bear two costs: the cost of allocating carbohydrates to berries at the expense of allocating them to other tissues such as leaves and the cost of losing leaves to herbivores. How blueberries and lingonberries, as well as other plants, resolve these opposing costs is a classic problem in plant ecology.

Given these differences between evergreen lingonberry and deciduous blueberry, it is reasonable to think that the two species will deal differently with the tradeoff between contributing carbohydrates to berry production and losing leaves to herbivores. To study how lingonberry and blueberry respond to these problems, Anne Tolvanen and Kari Laine of the Department of Biology, University of Oulu in Finland, conducted some simple and elegant experiments.[1] The experiments consisted of removing leaves and shoots (simulating herbivory by a munching animal) and removing flowers (stopping berry production). They then measured the growth responses and carbohydrate contents of the remaining leaves, shoots, and berries and compared those of the "munched" plants with those of the "unmunched" plants and the "deflowered plants," which didn't produce berries, with the plants whose flowers were intact, which did produce a crop of berries.

Tolvanen and Laine found that blueberry and lingonberry shoots and leaves indeed have opposite responses to flower removal. Lingonberries with intact flowers that yielded a crop of berries had slower-growing leaves and shoots than plants whose flowers had been removed. For lingonberries, berry production apparently comes at the expense of producing leaves and shoots. In contrast, blueberry shoots whose flowers and berries were left intact grew faster than those that had been deflowered, indicating that blueberries put their carbohydrates into shoots that are going to produce berries at the expense of the growth of other shoots without flowers.

The two species also responded differently to having their leaves removed, such as by a foraging moose. Blueberries that had their leaves and shoots removed regrew shoots to replace them, but the berries on those shoots had lower carbohydrate contents than those from unmunched plants. In contrast, the lingonberries that had their leaves and shoots removed grew fewer shoots and berries, but the berries on the regrown shoots had higher carbohydrate contents than those of unclipped shoots. In other words, munched blueberries sacrifice berry carbohydrate content to regrow berries, whereas munched lingonberries sacrifice shoot and leaf regrowth to increase carbohydrate content of the berries on the remaining stems. Tolvanen and Laine conclude, "The evergreen lingonberry grows slowly and conserves resources, whereas the deciduous blueberry allocates resources to increase the photosynthetic biomass."

Neither strategy appears to enable either blueberries or lingonberries to outcompete the other when they grow together, so one strategy isn't uniformly better than the other. Blueberries and lingonberries have simply evolved two different solutions to the same problem (how to keep growing and produce berries with seeds for the next generation) while facing the same set of constraints. The major constraint on both species appears to be the total amount of resources available from the soil, but there are several ways of maintaining their natural economy on limited resources. The shoots, leaves, and berries of the plant must cooperate, given the severe constraint of limited nutrient and carbohydrate resources. Although there are a number of ways for plant parts to cooperate, there are also a number of ways not to cooperate (for example, the plants apparently can't have both evergreen leaves and high photosynthetic rates).

The suites of traits in lingonberry (be an evergreen, grow slowly, and increase carbohydrate content when munched) and in blueberry (be deciduous, grow faster, and decrease carbohydrate content when munched) have evolved together within each species to facilitate survival of the population. But the evolution of these traits has produced two different solutions, which we differentiate as the blueberry or the lingonberry solution.

Both blueberries and lingonberries have highly organized systems representing differently engineered solutions to a design problem. The same design problems are faced by every other plant species. The nice thing about using blueberries and lingonberries to study these problems is that you can make pie or jam from your experimental material.

18.
Crossbills and Conifer Cones

Coevolution of conifer cones and the strange beak of the crossbills, the North Woods equivalent of Darwin's finches.

Think for a moment about birds' beaks. Think about the varieties of their forms and uses: the flat, rubbery bills of ducks and geese straining mud and water for algae and aquatic insects; the flesh-ripping hooks of eagles and hawks; the long and slender probes of marbled godwits jabbing the beach sand for mussels and worms; the chisel of the pileated woodpecker. Beaks are used for courtship, for grooming and preening, for building nests, for turning eggs, for capturing food and feeding young, for defense, for the many things a bird must do to be a bird. A bird meets and negotiates its environment largely by means of its beak. Because of this intimate association between a bird's beak and the environment it faces, mutation and natural selection have produced as many varieties of bird beaks as there are species, both extant and extinct.

The most famous examples of how natural selection modifies beak sizes and shapes are, of course, Darwin's finches on the Galápagos Islands. Darwin's finches are a drab group of species comprising four genera in the subfamily Geospizinae; they are distinguished from one another mainly by the sizes and shapes of their beaks, ranging from

197

small, warbler-like needles to large nutcrackers. Darwin himself did not recognize the significance of these birds when he visited the Galápagos, being more taken with the giant tortoises and marine iguanas. Because of the variety of their beaks, Darwin originally thought these finches were really a mix of wrens, orioles, and grosbeaks. But in March 1837, after he returned to England, he was astounded when John Gould of the British Museum of Natural History, to whom fell the task of describing Darwin's bird specimens, told him that they were different species of finches, thirteen in all, whose nearest relatives were on the South American mainland. Darwin was thunderstruck: How could this diversity of beaks among closely related species have come about? What was their relationship to other species 800 miles away over the sea? Thus, it was in a London drawing room, and not on the Galápagos Islands themselves, that the beaks of finches forced Darwin to face the problem of the origin of species.

But nobody, least of all Darwin himself, who was a modest fellow, called them Darwin's finches until almost exactly 100 years later when David Lack, an ornithologist whose career was dedicated to the evolution of the diversity of bird life, wrote a book called *Darwin's Finches*.[1] This book, even more than *The Voyage of the Beagle* or the *Origin of Species*, made the beaks of these finches iconic symbols of evolution. Lack showed that the sizes and shapes of beaks of the different finch species are adapted to the different sizes and hardness of the seeds from the particular plant species they eat. These adaptations for handling seeds from different plants and cracking them open enable Darwin's finches to avoid competition. Indeed, it was the accumulation of small adaptations in the beaks to different diets that divided populations one from another, thereby preventing competition, cross-breeding, and the sharing of genes, which eventually led to the radiation of the various species collectively known as Darwin's finches.[2]

In his book and in two short notes in *The Ibis*, Lack noted that the closest parallel to Darwin's finches are the crossbills, five species of the genus *Loxia* in the same family as finches.[3] The crossbills live in the conifer-rich northern half of the North Woods and in the boreal forests at the border between the United States and Canada, the conifers of the higher elevations of the Rocky and Cascade mountain ranges, and the long swath of conifer forest from Scotland through Scandinavia and Siberia. There are at least two species (and possibly more) of crossbills in North America: the red crossbill and the slightly larger white-winged crossbill. Crossbills breed in northern Minnesota along the Lake Superior shore and in the Adirondacks, Vermont, New Hampshire, and northern Maine.

The beak of the crossbill is one of the strangest of all bird beaks. The upper bill emerges straight from the skull in normal fashion, but the lower bill curves to the right or to the left before turning its tip upward. The two tips do not meet but instead cross to one side or another. The beak looks like a broken set of pliers.

But therein lies its strength. Crossbills feed exclusively on conifer seeds, but to get those seeds they must pry apart the armored protection of the conifer cone. The crossbill does this by holding the cone transversely between its feet, as if the cone were a corn cob. The cone is held by the foot on the same side of the bird as the lower bill curves. Whereas we are right- or left-handed, the crossbill is right- or left-beaked and right- or left-footed.

The crossbill next inserts the straight upper bill between two adjacent scales of the cone and lays its side against the flat bottom of the uppermost scale. It then places the bent tip of the lower bill against the lower scale and, by opening its jaw, forces the two scales apart. This maneuver exposes the seed, which is held in a pocket at the base of the lower scale. Finally, grasping the seed with the spoon-shaped tip of its long tongue, the crossbill holds it in a groove in its palate and removes the husk before swallowing the energy-rich kernel.

The crossbill continues in this way to extract all the seeds from the cone, much as we eat corn on the cob, flinging it to the forest floor after the cone is empty. My wife and I got a spectacular demonstration of this the morning of Christmas Eve 2013, when several dozen white-winged crossbills in a grove of large white spruce in our front yard were flinging and raining cones down to the snow and on the sidewalk. After they were finished, I swept up two 5-gallon pails of cones from the sidewalk, which I used to light fires in the fireplace the rest of that winter and the next.

Crossbills stay together in flocks of several dozen and don't migrate far except in years when the cone crop in the north has failed. In those years, their populations then irrupt, or spread, a bit farther south. The flock in our yard on Christmas Eve was probably part of an **irruption** from farther north in Minnesota or Ontario.

Because they do not migrate far and because flocks stay largely intact, interbreeding and therefore gene flow between isolated populations of crossbills is low. A mutation that appears in one population of crossbills may thereby be preserved and spread down through successive generations in that population if it promotes the survival and number of offspring, but it will probably not spread to other populations from which it is isolated. Genetic isolation is a prime condition for preservation of different mutations in different populations. Eventually, generation after generation, the appearances and behaviors of the different populations diverge. Given enough time, the two populations may become sufficiently different to prevent interbreeding even if they came together. Somewhere along this divergence they would have become different subspecies and eventually different species.

This process of adaptive radiation is how the different beak shapes in Darwin's finches probably evolved, and something of the sort is also happening in different subspecies of crossbills. Because the sizes and shapes of beaks determine which foods Darwin's finches and crossbills can eat, they are the key mutations that genetically isolate populations one from another. For crossbills, the two main properties of their beaks that determine foraging efficiency are the depth of the beak and the width of the groove in the palate where the seed is held while being husked. In a series of articles on the natural history of the crossbill, Craig Benkman has shown that the depth of the beak controls the time a crossbill takes to remove seeds from cones, and the width of the palate groove determines how long the crossbill handles the seed to remove the husk before it can swallow it.[4] Benkman also determined that beak depth is highly heritable, meaning that birds with deep beaks have offspring with deep beaks, and birds with shallow beaks have offspring with shallow beaks. Presumably, palate groove width is also heritable in the same way.

There is an optimum combination of beak depth and palate groove width that minimizes the time a crossbill takes to extract and process seeds from the cones of their conifer prey. Crossbills with these optimum beak shapes and palate grooves can eat more seeds (and thereby gain more energy) in a given hour of foraging than crossbills whose beak shapes and groove widths depart from the optimum dimensions. The more seeds obtained per time, greater the chance the crossbill will survive the winter and therefore get a chance to mate and pass on these beak characteristics to its offspring the next spring. Even small differences in beak depth and groove width can make large differences in survival: Half a millimeter of beak depth saves 1 second in the time it takes to pry each seed out of a lodgepole pine cone for red crossbills in the pine forests of the South Hills in Idaho.[5] Half a millimeter of beak depth and 1 second of extraction time per seed makes a huge difference in reducing the time it takes a crossbill to obtain the many tens of thousands of seeds needed to supply the energy to get through winter. (To see how small half a millimeter is, look at a spruce needle; these are about a millimeter wide.) This half millimeter can mean life or death. Crossbills in Idaho with a beak depth of 9.5 millimeters have less than a 20 percent chance of surviving at least 1 year, but those with a beak depth of 10 millimeters have a better than 60 percent chance of survival.

The optimum beak depths and groove widths vary with the species of cone the crossbill is dismembering. Red crossbills with 8-millimeter beak depths and 1.85-millimeter groove widths are most efficient at extracting seeds from western hemlock in the northern Rocky Mountains and Cascade Mountains of the Pacific Northwest.[6] On the other hand, crossbills with 9.6-millimeter beak depths and 2.13-millimeter groove widths are most efficient at extracting seeds from ponderosa pine. Western hemlock and ponderosa pine rarely grow together, so these two crossbill populations, each specializing on the cones from these different conifer species growing in different habitats, are isolated from interbreeding. The exchange of genes between these two populations is inhibited by these small differences in beak sizes. Benkman thinks these populations are sufficiently isolated in habitat and feeding behavior that they constitute different subspecies, if not different species entirely.

But the cones can fight back, so to speak. Cones with thicker scales are more difficult for crossbills to handle, especially scales at the distal end of the cone (away from the twig), where crossbills prefer to feed.

Trees that produce such cones therefore have more seeds that escape being eaten by crossbills. The genes for these cone characteristics then begin to increase in successive generations of trees. This appears to have resulted in a variety, if not a subspecies, of black spruce in Newfoundland, where crossbills until recently have been the main consumer of conifer seeds.[7]

In contrast, on the mainland of eastern Canada, red squirrels, rather than crossbills, are the main consumer of black spruce seeds.[8] Red squirrels, unlike crossbills, simply bite through the cone to get the seeds, especially at the basal end near the twig rather than at the distal end preferred by crossbills. Cones of black spruce on the mainland, which are wider and have thicker scales at the basal end, can therefore more successfully protect themselves against predation by red squirrels because the squirrels have to bite through a larger and tougher scale mass to get to the seeds at the basal end. The spruces on the mainland and on Newfoundland have become adapted in different ways to protect themselves against predation by either crossbills or red squirrels, whichever is the more prevalent consumer. Similar divergent responses of other conifers, such as lodgepole pine and other western conifers, have also happened depending on whether crossbills or red squirrels are the major predators of those cones.[9] This evolutionary arms race between conifer cones and their crossbill and squirrel predators is partly responsible for the diversity of crossbill and conifer subspecies across the North Woods and conifer forests of northern North America.

Red squirrels were initially absent in Newfoundland, and the crossbills had the black spruce all to themselves.[10] The isolation of the Newfoundland crossbills produced a unique subspecies known as the Newfoundland red crossbill. Newfoundland red crossbills evolved deeper beaks than crossbills on the mainland, who faced strong competition for cones from red squirrels.

Unfortunately, the Newfoundland red crossbill may be on the way to extinction. In 1963 and 1964, the Newfoundland Wildlife Service introduced red squirrels to Newfoundland to provide prey to assist the recovery of pine marten, whose populations had been driven dangerously low by overtrapping. Because red squirrels outcompete crossbills for spruce cones, the squirrels rapidly became the stronger selection force on the size and shape of spruce cones. The spruce cones are now evolving wider and thicker scales to protect themselves against the introduced red squirrels, but the native Newfoundland crossbills are finding it difficult to extract seeds from these cones. Consequently, the population of Newfoundland red crossbills has declined precipitously. The irony is that the introduction of red squirrels to assist in the recovery of the declining pine marten seems to have contributed to the decline of the Newfoundland red crossbill.

Populations and species evolve, but the direction of evolution is always within the context of the rest of the food web. Without a sound knowledge of the natural history of a species, it is difficult to predict which way evolution will take it. Even when we do have a sound knowledge of natural history, evolution often throws us a curveball. It often doesn't take much difference in a trait to reduce population viability. As Benkman has shown, a half-millimeter difference in beak depth is sufficient to substantially change survival rates of crossbills. Our conservation efforts, even the most well meaning, may fail when they do not take natural history and evolution into account. Natural history and evolution therefore underlie all of conservation ecology and resource management. The two North American species of crossbills, white-winged and red, are not themselves in danger of extinction across their ranges, but some subspecies within these two species seem to be in decline, ironically partly because of laudable efforts to conserve other species.

PART V
Fire and the Dynamics of the Landscape

For more than a century, fire was considered the greatest catastrophe that could befall the North Woods or any other forest. Then, research in the 1960s began to suggest that, for the past 500 years at least, every stand in North Woods west of Lake Superior originated from fire. This and other research led to the conclusion that fire is an important component of the North Woods ecosystem and is one of the forces responsible for maintaining its species composition, structure, and overall character.

But if fire is necessary to maintain a forest, then how do forests recover from fire? Part of the answer lies in the trees that survive a fire, the charred snags that do not, and other legacies of the burn. These legacies are reservoirs of DNA, energy, and nutrients that enable the next forest and its food web to rise from the ashes of its predecessor like the mythical Phoenix.

If all this is so, then fire must be a strong selection pressure on species, especially in their reproduction after the fire. The cones of jack pine and to some extent black spruce need fire to melt the resins that hold them shut and disperse their seeds, a trait known as **serotiny**. But jack pine and black spruce needles are among the most flammable of all conifer needles. Could flammability and serotiny have coevolved? If so, then evolution has unleashed a force that shapes immense portions of the northern landscape.

19.

Does Fire Destroy or Maintain the North Woods?

The distribution of fire in the North Woods, the variety of adaptations to it, and the diversity of a fire-dominated landscape.

Sometime in early- to mid-August 2011, lightning struck a black spruce snag in a small bog in the Boundary Waters Canoe Area Wilderness in northern Minnesota, near Pagami Creek. The fire lingered for several days in a few hundred square meters of the bog, smoldering in the damp peat, before smoke was detected on August 18. Because the fire seemed confined to a small portion of the bog, and because natural fires are allowed to burn in the Boundary Waters, the Forest Service decided to monitor rather than fight it. Then, on August 26, the relative humidity plummeted to 18 percent and a strong wind blew from the north. The fire was whipped into the crowns of other spruce trees and out of the bog, headed due south. By the end of the day, the fire had a north–south flank 2½ kilometers long and had burned half of a square kilometer. By September 9, the same north winds had expanded the fire to 18 square kilometers.

Then, in the morning of September 12, the wind shifted direction and began to blow from west to east. The whole north–south flank of the fire, now several more kilometers long, was blown 13 kilometers east by late morning and another 9 kilometers by late afternoon. Now the fire had a southern flank that was 22 kilometers long. A wedge of white smoke with its apex at the ignition point and its base over Lake Superior was visible from space.[1]

By the early evening, the wind shifted to the northwest and blew the entire southern flank 16 kilometers south. By the time the sun set, nearly 376 square kilometers, mostly old growth pines, had burned. The plume from the fire, containing about 25,000 kilograms of soot, was carried aloft to the upper troposphere, where winds blew it east across Canada, the Atlantic Ocean, and into Europe.[2] Within a few days, two burnt fingers pointed southeast out of the wilderness area, and Forest Service crews began to successfully contain the fire, lest it send the nearby logging town of Isabella up in smoke along with the forest.

The fire left a patchwork of bare rock where the fire had been so hot it burned away the topsoil, stretches of standing dead and charred snags, and pockets of forests that escaped ignition.

How should we think about this and other fires? Did the fire destroy the forest and homogenize the landscape to a single age class of trees that followed? Or was the fire an essential process for maintaining the diversity of forests?[3] For many of us raised on a steady diet of Smokey the Bear saying, "Only You Can Prevent Forest Fires," implying that they must be prevented at all cost, the thought of fire as a necessary part of a forested ecosystem makes no sense. A fire burns, and the forest is destroyed. But we can also think, "The fuels in a forest burn, and the forest is renewed." That we can even seriously consider fire to be an integral part of a forest ecosystem rather than its destroyer is due in large part to the pioneering work of Miron Heinselman, known to many of us simply as "Bud."

Bud was born in Duluth, Minnesota in 1920 and was a research scientist in St. Paul with the U.S. Forest Service Lake States Experiment Station (now Northern Research Station) from 1948 until 1974. He did not publish many articles, but the ones he did were monumental and laid the foundations of both peatland ecology[4] and fire ecology[5] (a lesson for faculty hiring and tenure committees). Bud's work on fire ecology was done almost entirely in the Boundary Waters Canoe Area Wilderness.

The Boundary Waters is the northern tier of the Superior National Forest in northern Minnesota on the border with Canada. The border lakes were the main canoe route inland from Lake Superior during the fur trade. Established in 1909, the Superior National Forest already was known for its large, old growth pines. By 1929, the general area of the current Boundary Waters was established as the Superior Roadless Area, a de facto wilderness, although some logging was still allowed around the edge. The Boundary Waters was named in the 1964 Wilderness Act, and the final boundaries, entry points, and lakes were established in the 1978 Boundary Waters Canoe Area Wilderness Act. Bud's maps of the virgin (uncut) forests of the Boundary Waters were essential material used by Congress to define the modern idea of wilderness in these two historic acts. The Boundary Waters, along with the Quetico Provincial Park to the north in Canada and Voyageurs National Park to the west in Minnesota, make up what is now called the Quetico–Superior Wilderness. At more than 10,000 square kilometers, the Quetico–Superior is the largest contiguous wilderness area in North America south of the Arctic. Between the Quetico–Superior and its twin, the state-owned lands in the Adirondack Park in New York, the North Woods has more land in wilderness than any other biome in North America except for tundra.

Natural forest fires can range from a few hectares to hundreds of square kilometers. When a landscape is in danger of burning, such as during a prolonged drought or when a sufficient fuel load has accumulated, many thousands of square kilometers can burn in a single year, as Bud discovered. The study of the natural history and ecology of fire therefore requires access to large tracts of wilderness such as the Quetico–Superior, which is mostly "untrammeled by man, where man himself is a visitor who does not remain."[6] Unfortunately, large tracts such as the Quetico–Superior are rare. As Aldo Leopold wrote, the primary value of wilderness areas is not for recreation but for land laboratories in which the dynamics of nature can be studied on the large scales over which they happen.[7] Before we can manage the natural resources of an area effectively, we need to know how its natural history and ecology work without interference by humans. Leopold said that we need the wilderness to teach us how ecosystems have sustained themselves as they have for millennia. We know that if the wilderness is large enough to include all its essential processes within its boundaries, then it will sustain itself without assistance from us; a forest does not need forest managers to be sustainable.

The Quetico–Superior Wilderness is one of the few places left in North America that qualifies as a true ecological wilderness, where fires can be studied on the enormous scales over which they burn, regenerate, and sustain certain types of forests. Having a large wilderness as your research site is one thing, but getting around in it is another matter entirely, especially because motorized travel in wilderness areas is prohibited by law. Motorized travel would not do much good in the Quetico–Superior anyhow, because the landscape is a bazillion lakes and ponds connected by portages, which are often long slogs over ridges or through bogs. Who would want to portage a motorboat through this landscape? In studying the forests of the Boundary Waters, Bud resorted to the proven method of travel through it, by canoe.

In late October 1991, I was invited to participate in a fire symposium organized by the Friends of the Boundary Waters Wilderness. The purpose of this symposium was to tour some of the oldest and best examples of virgin forests with Bud (yes, by canoe) and learn how he deciphered the fire history of this region. From our canoes, Bud pointed out where the appearance of the canopy of the predominantly white and red pine forests changed abruptly at various points along the shore. This was a clue that the stands of different canopies on either side of this boundary dated from different times. Younger forests have largely uniform canopies because they are composed of trees of approximately the same age, which began their life together in a previous burn. As the forest ages, some trees die and others grow into the gaps they left. Branches on the larger and older trees are broken by multiple insults of ice, snow, and wind. Dominant trees continue to grow and tower over younger trees, shed their lower branches, and become more irregular. The individual crowns and the entire canopy assume a rougher appearance as the older trees loose some of their branches and as the forest becomes composed of both young trees growing into gaps and older trees towering above the rest.

With a little practice, we could easily identify forests of potentially different ages while in the canoe. During his research, once Bud identified what appeared to be forests of different ages, he then beached his canoe on shore and sought out one or more of the largest pines, preferably a red pine with a fire scar. Flames swirl around the downwind side of a trunk during a fire and leave a triangular scar (see Essay 20 for more details). The base of the triangle is at the base of the tree because the fire is hottest near the forest floor. As the heat from the fire rises, it cools and burns a progressively narrower scar up from the base, eventually forming the apex of the triangular scar. Within the scar, the bark and the living **cambium** of the wood, which is only a few rings wide, are often killed by the fire, but the cambium around the rest of the circumference of the trunk is unharmed. This unharmed cambium begins to grow laterally into the scar, making wood toward the interior of the trunk and bark toward the exterior. The scar may remain for many decades as a triangle of dead wood recessed into the growing wood and bark, but sometimes the scar heals over completely from the two sides, leaving a telltale flat, triangular shape on the otherwise rounded trunk.

Near the base of the tree, Bud used an increment borer to take a core from the unharmed portion of the trunk of the tree opposite the scar. This gave him a complete record of rings from which he could determine the age of the tree. To learn when particular fires happened in the past, he cut a small wedge from the fire scar and adjacent unburned trunk with a handsaw (no chainsaws in the wilderness!), making first a horizontal cut into the trunk and then a downward slanting cut meeting the first cut some distance in from the scar.[8] Surprisingly, taking these cores and wedges does not kill the tree. Successive fires during the tree's life left blackened rings in the wedge. By counting back from the most current ring, Bud could obtain a fire history of the stand, often going back several centuries for very old sentinels of the forest. Many of these trees recorded several fires during their lifetimes, each fire leaving a blackened ring separated by normal ring growth.

Once he learned the age of a stand and its fire history from these cores and wedges, Bud could calibrate the stand age and its fire history to the appearance of the canopy, which he had noted previously while in the canoe. He then made field maps of each stand from the canoe, noting its age and the years it burned. These field maps were later supplemented with data from air photos back in the office and transferred to U.S. Geological Survey topographic quadrangles. Bud then superimposed a transparent sheet with a grid of dots over each map. By counting the number of dots in each stand type, Bud could calculate the proportion of the total area of the quadrangle burned in each fire. Bud scribbled these dot counts and calculations from them in the right-hand margins of the topographic maps along with many field notes, comments, and questions.[9]

Nowadays, we would measure the areas of each burn using a geographic information system (GIS). This is more precise and probably more accurate. But as I was looking recently through these hand-drawn maps, sixty in all, at the John R. Borchert Map Library of the University of Minnesota, I began to wonder whether Bud's ideas and theories about the importance of fire in these ecosystems dawned on him as he plodded through the painstaking efforts of tabulating these dot counts, map after map. Counting dots, making calculations, and feeling the shape of the stand boundaries as he drew them by hand while thinking about what he saw from his canoe must have given Bud a deep feeling for the behavior of these fires and how they created the landscape; no GIS could provide such an intuitive feel for the landscape in quite the same way. At some point during these tasks, Bud must have had his *Eureka!* moment when he realized that every virgin, old growth forest in the Boundary Waters alive today contains at least a few trees, some of them centuries old, that had one or more fire scars. This means that every one of today's virgin forests began in burns of previous old growth forests, and many of these virgin forests burned several times. This finding changed the way we think about forest fires. Previous fires did not simply destroy the previous generations of forests; they were responsible for the birth of every stand we see today.

The oldest pines in the Boundary Waters were a small grove of three red pines dating from 1595 (according to the core Bud took from one of them) on Threemile Island in Seagull Lake, in the Munker Island Quadrangle. These pines had fire scars from 1692 and 1801, which were evident in the wedge Bud took from one of them, but there were no fire scars in the wedge later than that. There is a small, unnamed island northwest of Threemile Island that also burned in 1692 and 1801. Next to it on the map is a pencil note in Bud's handwriting with this query: "Many very large RP–WP [red pine and white pine]. Some could be 1595?" Despite Bud's note to himself that some of these trees on this unnamed island could also date from 1595, I don't think he ever got the opportunity to revisit and take cores from them.

By October 20, 1991, when we visited the 1595 grove on Threemile Island with Bud, two of the three 1595 red pines had blown down. True to form, Bud later added this note to the map next to Threemile Island:

"I visited the 1595 origin Red Pine 10-20-91 with the FOBWW Fire Seminar group. One tree still standing—*it is the one I got the age from in 1969!* Photos taken."[10] Proof that taking a wedge and core from a tree does not kill it.

These maps also document the scale and timing of major fires in this region. Somewhere in the Boundary Waters, there was a fire significant enough to leave fire scars every 4 years or so. Most of these fires were small, but a few years with enormous fires even larger than the Pagami Creek fire account for 83 percent of the total area that burned between 1600 and 1900. These fires happened in 1681, 1692, 1727, 1755–1759, 1801, 1824, 1863–1864, 1875, and 1894, approximately every 26 years. Of these years, 1894, 1875, and 1863–1864 account for 23, 22, and 20 percent, respectively, of the virgin forests that burned. The 1863–1864 fires were almost completely unknown before Bud's dating of them from the fire scars, because the country was then preoccupied with Sherman's March through Georgia, with its attendant burning of crops and farms. Smoke hung heavily in the air across the country during those years.

The average amount of time between major fires somewhere in the Boundary Waters is not the same as the amount of time between recurring fires in a given stand, sometimes known as a fire return interval or, in Bud's more descriptive term, the natural fire rotation. On average, any single Boundary Waters forest has a natural fire rotation of about every 100 years.[11]

Although once a century may be the average natural fire rotation across the entire Boundary Waters Wilderness, the actual fire rotation varies between forest types and depends on the flammability of the fuel and the water holding capacity of the soil on the different glacial deposits.[12] Some aspen–birch stands on sandy outwash or thin and rocky soils may burn every 50 to 100 years or so. These frequent fires maintain aspen and birch: Without the fire and when they reached the end of their lives (at about 80–100 years of age), the aspen and birch would have eventually been replaced by individuals of other more shade tolerant species, such as maple, spruce, or white pine, growing in the understory. Although the fires probably killed most of the aspen and birch stems aboveground, they were probably not hot enough to kill the roots of aspen or the stump of birch. Aspen roots and birch stumps promptly send up new shoots when the main stems die; these shoots eventually grow into a new forest of mature aspen and birch. You can recognize a tree of this second generation of birch because many of the trees have multiple trunks originating from the same base; these are the shoots from the stump that survived. If another fire comes along and burns the mature aspen or birch forest, the process repeats itself.

These frequent fires also open up the cones of jack pine, which are glued shut by resins and scatter many years' worth of seeds on the ground. These seeds then germinate and regenerate a new jack pine stand (see Essay 21). Indeed, without frequent fires to clear the overstory and open cones, jack pine would not have been as abundant in the virgin North Woods of the Lake Superior region as it was.

Red and white pine stands have a more complex fire regime, consisting of severe crown and ground fires in intervals of 120 to 180 years punctuated by more frequent but lighter ground fires every 30 to 50 years or so. Fires can roar through their forest floors with ease, partly because of the high resin contents of the needles, partly because the needles form a thick and fluffy layer containing much air, like piles of children's pick-up sticks, and partly because the slow decay rate of conifer needles allows many annual cohorts of needle litter to accumulate. These three properties of pine forest floors create a thick, dry, resinous layer of old needles that are highly flammable.

This complex, two-stage fire regime of frequent ground fires and infrequent crown fires favors red and white pine. These pines, especially red pine, have very thick bark at their bases that can protect the larger trees from both light and severe ground fires. But ground fires would have killed any competing understory vegetation, especially individuals of thin-barked species such as maples, juneberries, viburnums, and other hardwoods. If the fire leaped to the crown, many old red and white pines would also have been killed. Unlike those of jack pine, cones of red and white pine are not glued shut and do not need fire to open. Any surviving old red and white pines continued to produce cones after the fire. When these cones opened during the winter after they were formed, their seeds were dispersed and became the sources for the next generation of trees, which became established in the now cleared understory. Survival of a few old seed trees and periodic clearing of the understory that made way for a new generation of pine seedlings were probably the only ways red and white pine populations sustained themselves in virgin forests.

Northern hardwood forests growing on clay-rich soils of moraines have a very long fire rotation, more than 300 years. This is partly because of the greater moisture content of these soils, which are derived from the clays and silts in glacial moraines rather than the sand on the outwash plains, where the drought-tolerant pines are more abundant (see Essay 2). Unlike conifer forest floors, the flat leaves of maple, yellow birch, basswood, and beech lie in imbricated layers in hardwood forest floors, overlapping each other like shingles on a roof. Films of water are trapped between these layers of flat leaves, keeping them moist. The high moisture trapped by these leaf litters, and their low resin and lignin concentrations, promote rapid decay so fuels do not accumulate in these forest floors as in conifer forests. Fires cannot get a purchase in these forest floors except during times of prolonged drought and abnormally warm weather. As a result, northern hardwoods are often known as asbestos forests.[13]

Fire is a more important component of the North Woods at the western end of its range in the Lake Superior region, where the climate is drier and conifers are more abundant, than in the east, especially in the Adirondacks, New England, and the Maritime Provinces, where rainfall is more abundant and fire is less common. The greater preponderance of hardwoods and their thinner forest floors, which hold water, and the greater rainfall are probably what give the North Woods its asbestos character in the east. The natural fire rotation in the east is greater than the expected lifetimes of the maple and beech and even the white pines. These species can live for four centuries or more. Because of the long recurrence interval between fires even in conifer forests of northern Maine,[14] most forests attain a steady-state composition of species and biomass, unlike in Minnesota, where fires are more frequent. In these eastern forests, fires seem to be a rare disturbance that happens once every two or three lifetimes of the trees. In contrast, fire in the western Lake Superior region seems to be a more frequently recurring and intrinsic process that maintains the forests in a diverse mosaic of early successional forests composed of shade-intolerant aspen, birch, and jack pine, midsuccessional forests of white and red pine, and late successional forests of maple and other hardwoods. Fire destroys the maple forests at the North Woods' wetter and eastern end, but fire is also necessary to maintain and rejuvenate the aspen, birch, and pine forests at the North Woods' drier western end. Therefore, the answer to the question of whether a fire destroys or maintains a forest depends on which type of forest you are considering. Posing it as an either–or question is a false dichotomy. Ecology is full of such false dichotomies, which can almost always be resolved by taking a larger view of the world, in this case by examining the roles of fire at both ends of the North Woods biome rather than focusing on any single forest in only one part of the North Woods.

In the evening between the 2 days of our field trip with Bud, we discussed some of the questions Bud's research prompted that remained unanswered (and remain so still). The most important question, which Bud said he could not answer and in some ways still did not understand, was "How does a fire spread through a landscape that previous fires have created?" This question has some affinity with the question we asked about moose: "How does a moose move through and forage in a landscape that its forebears have created?" (see Essay 11). Both burning and foraging are spatial processes that consume the fuels or forage that sustain them. Fires and moose populations also alter the future distribution of fuel and forage and create the templates for the spread of future fires and moose populations. We have seen how moose depend to some extent on the regeneration of aspen by fires and how selective foraging by moose increases the dominance of flammable species such as fir and spruce, so moose foraging and fire are mutually dependent to some extent. Recent developments in landscape models that incorporate both life history attributes of component species and spatial patterns of soils and topography may prove to be promising tools to answer these questions.[15]

Bud always wanted to see one of the colossal natural wildfires, such as those that burned in 1863–1864, but unfortunately he died in 1993 and so did not live to see the Pagami Creek fire in 2011. When I taught my ecosystems ecology class that fall, I decided to give a few lectures about natural fires using the Pagami Creek fire as an example, especially because several of the students had been on the fire crews that fought it. Wondering how long it had been since there was a major fire in the Pagami Creek burn, I consulted the maps in Bud's 1973 article, which indicated that most of the area that was burned had not had a fire since 1863–1864, or 147 years earlier. Some of this area had been logged in the first decades of the twentieth century and now supported 80-year-old jack pine and black spruce; much of the remainder was old growth red and white pine stands that began after the 1863–1864 fire.[16] A major fire after 147 years for old growth red and white pine and after 80 years for jack pine and black spruce is right in the middle of the natural fire rotations for these types of forests. Bud would have said that the Pagami Creek fire arrived right on schedule.

20.

The Legacies of a Fire

What a fire leaves behind determines how the next forest grows and how the new food web develops.

Poet and biologist Miller Williams wrote, "We are surrounded by ghosts. Reminders of past and nearly forgotten days are all around us, things neither alive nor dead from an old world. We see them, but we may not notice."[1] The ghosts of past fires—the scarred trees, still alive and waiting to disperse the seeds of the next generation; the burnt and charred upright snags and downed logs; the pockets of unburnt forest; the bare rock where the soil has been burnt or eroded away and where nothing will grow for many decades, even a century; the seed bank in unburned humus—are **legacies** handed down from one forest to its successor. These legacies are links in a long chain of forests, providing continuity from one forest to the next.[2] Legacies record the presence and behavior of past fires, bequeath the genetic information in seeds that will spark the growth of the next generation, and provide structures on which the new forest and food web can rebuild. It is only in the past decade or two that we have begun to notice legacies and begun to understand how they link past, current, and future forests into a continuing ecosystem.

We saw in Essay 19 how Bud Heinselman reconstructed the fire history of the Boundary Waters landscape from the blackened rings in trees that survived the fires. As part of his research reconstructing the fire history of the Boundary Waters Canoe Area Wilderness, Bud Heinselman also visited recent fires to see how the scars were actually made in the hopes that this knowledge would help him better interpret the nature of past fires. He found something striking: Fire scars are almost always found on the downwind side of trees because the scar forms in the vortex of hot air and flames circulating in the lee of the large trunk. Fire scars record not only the history of when fires happened but how fire moved through the forest. Of course, this is difficult if not impossible to document while a fire is raging. But by standing at the scar, facing away from the tree, and measuring the compass direction the scar was facing, Bud knew that had he been there at the time of the fire, it would have been roaring down his back.

I decided to try my hand at this one day in a pine stand near Little Trout Lake, in the eastern portion of Voyageurs National Park in northern Minnesota. Little Trout Lake is separated from Sand Point Lake to the south by a bedrock ridge. The ridge is about half a kilometer wide and rises about 20 meters from either lake shore to its crest. It trends almost exactly northwest–southeast. Its northeastern slope faces Little Trout Lake, and its southwestern slope faces Sand Point Lake.

There are many large white and red pines along the short trail over this ridge. Although I didn't have a borer there to take cores and count the rings, from their large diameters (many between three-quarters of a meter to more than a meter), I am nearly certain that the largest of these trees were more than 100 years old. Quite a few had fire scars. The trees that are greater than 50 centimeters in diameter almost invariably have at least one fire scar, whereas smaller trees usually don't have any. This means that two generations of trees were present in this stand: one generation represented by the larger scarred trees that had survived at least one previous fire and a younger generation of smaller trees without fire scars, which presumably had grown after the last fire.

While wandering through this stand, I noticed that the fire scars faced different directions depending on where on the ridge the tree was. Taking out my compass, I started to record directions and make a sketch map in a notebook. On the northeast side of the ridge, toward Little Trout Lake, the fire scars were small and indistinct. They faced almost due west, away from the lake. On the ridgetop, some of the scars faced more northwestward, directly along the ridgeline. Going over the ridge and down the southwest side toward Sand Point Lake, the scars were larger and were arranged in two different paths, one whose scars faced northwest and another whose scars faced southeast.

My impression was that these scars were made by a single fire because they were all healed over to the same extent. The data on the compass directions they faced is the record of how this fire moved through this stand. It seems to have come to the ridge from the eastern or southern shores of Little Trout Lake. It then burned rapidly westward and uphill, so rapidly that there was little time to form scars on the northeast side of the ridge, hence their small and indistinct nature there. Once it reached the ridgetop, it then traveled northwest along it for a small distance. At this point, it began to burn downhill but split into two branches, one continuing northwest, the other heading southeast. Why the fire split into two branches is not clear, but the updraft of hot air within a fire interacts with the topography in complex ways and creates its own weather inside the burn. This is part of the reason why even veteran fire crews can be trapped suddenly between prongs of fires, often with tragic consequences. Burning downhill is a slow process, and the fire here seems to have had more time to scar the trees, creating the much larger scars compared with the smaller scars on the other side of the ridge. The pines that survived this fire are the ones that are now more than 50 centimeters in diameter. The pines smaller than 50 centimeters without scars appear not to have seen a fire and are apparently the prog-eny of the survivors.

We can't know when this fire happened without taking cores and wedges from the trees and counting rings (see Essay 19 for an explana-tion of these techniques). However, we know from Bud's work in the Boundary Waters that there was a large fire in 1864 near the southwest-ern shores of Lac La Croix, just to the east of Sand Point Lake. We also know that a large white and red pine stand at the mouth of Mica Bay of Lake Kabetogama, west of Little Trout Lake, partly burned in 1864. In fact, the pattern of fire-scarred large trees and unscarred smaller trees in the Little Trout Lake stand fits the description of the Mica Bay stand and some of the La Croix stands. Little Trout Lake sits right smack in the middle of a straight line from La Croix to Mica Bay. I'm willing to bet that a huge fire started somewhere around La Croix and burned westward through Little Trout Lake onward to Mica Bay in 1864. The western extent of this fire is unknown, but there are plenty of pines west of Mica Bay waiting for someone to learn from them.

The pine stand at the mouth of Mica Bay provides another interest-ing example of the legacies fires leave. We discovered this stand on the Kabetogama Peninsula when we were studying the ecology of beaver ponds. While using historical air photos to estimate beaver population trends, we noticed that a pine stand at the mouth of Mica Bay was intact going back to the early 1940s. We later obtained a set of air photos from 1924 that the Royal Canadian Mounted Police took (from biplanes!).[3] On these photographs, we saw logging slash on both sides of Mica Bay, with the logs being stored in the bay by a boom chain crossing the mouth. But the pine stand just east of the mouth was uncut. Logging at the time was a winter operation, and logging outfits often did not return the next winter to finish the cutting. That is apparently how this stand was spared, but no one, including the Park Service, knew of its existence until we saw it on these old photos.

Thinking that this pine stand east of the logging slash might be an old growth stand, I took our boat up there one day to take a look. It was (and still is) a superb old growth white and red pine stand of about 100 hectares. This is the largest old growth stand in Minnesota outside the Boundary Waters Wilderness. It contains the largest white pine I know of in Voyageurs National Park (and one of the largest I've seen anywhere), a magnificent stalwart with a diameter of 120 centimeters. I defy anyone not to hug this tree and say, "Well done, old man." This white pine had a large fire scar at its base. This didn't surprise me, but as I looked around for others I could see that red and white pines of 90 centimeters diameter or greater had fire scars, but smaller trees of about 50 to 70 centimeters did not. These smaller trees must be the progeny of the larger trees that survived the fire. As I mentioned previously, the fire probably happened in 1864, so the younger trees themselves qualify as old growth, which in Minnesota is a forest that has not had a major, canopy-opening disturbance for at least 120 years. The large, fire-scarred veterans are the legacy of the 1864 fire that gave birth to the present stand, which survived logging by the merest chance.

The seed rain from a tree is densest closest to the tree and falls away exponentially with distance, so for some time after a large disturbance such as a fire, the seed trees create legacies of patches of the same species surrounding them.[4] But as I walked through this stand I began to wonder whether these patches remained intact into the next old growth stand or whether random deaths of the next generation destroyed the patterns over time. Because there seemed to be only two generations of pines in this stand—the fire-scarred, large veterans that survived the fire and became the legacy seed trees and the younger (but now old) pines without fire scars—this stand seemed an ideal choice to answer this question.

So, a few days later I returned with Cal Harth, one of my technicians, and we began to inventory the distribution of trees in this stand. First, we located a dozen each of the old fire-scarred white pine and red pine veterans and designated them as centers of 0.1-hectare plots. Within these plots, we counted and measured the diameter and noted the species of all trees greater than 5 centimeters. We then located another dozen 0.1-hectare plots randomly throughout the stand. If the legacy of the veteran seed trees persisted into the next old growth stand, then there should be a greater number and basal area of white pines or red pines in the plots centered on the veteran seed trees of these two species that survived the fire than in the plots distributed at random.

But we found no difference in species composition between the plots centered on the veteran pines of either species and the plots distributed at random. Apparently, the spatial patterns of the next generation of progeny of the surviving seed trees did not persist into the next old growth stand, at least in this case. This was a surprise to me. Legacies have a long lifetime (that's why they are called legacies), but they also decay with time. The rates of decay of legacies are perhaps their least known (and least well understood) properties. Maybe with more plots we might have detected a statistically significant difference (more plots and bigger plots are always the answers to our field problems). But if so, then the effect probably would have been a weak one subject to high variance in the data. At the very least, the influence of the seed trees on the spatial pattern of the next generation has become severely weakened by the time their progeny has grown to become the next old growth stand. I never published this because the results were negative, and journals and reviewers are notoriously reluctant to publish negative results (a negative result is one in which the null hypothesis of no effect is upheld). Offering these results here might encourage a graduate student to take this problem on as a thesis topic.

Whereas the live survivors of a fire provide the link between generations, the legacies of charred and blackened dead snags are important structures for the reassembly of the post-fir food web. There are a host of insects of northern forests, mainly wood burrowing and bark beetles, that are known as fire-dependent species. Of course, these insects don't depend on the fire itself but on the charred snags, in which they burrow and lay their eggs. Populations of these species erupt rapidly after a fire for a few years, then promptly decline. But in those few years, they form the base of the recovering food web. Many of these species are rare and endangered because of decades of fire suppression.[5] The decreased populations of these beetles may affect other species in the food web. In the North Woods and boreal forest of North America, one of their most important predators is the black-backed or Arctic three-toed woodpecker, a species that is also now rare because it is a feeding specialist on these fire-dependent beetles.[6] Scandinavian ecologists have long emphasized the importance of charred snags in maintaining insect biodiversity of boreal forests, but in North America we have barely begun to understand how burned and charred snags provide food and habitat for the species that make up the succeeding food webs after fires in the North Woods or elsewhere.

Legacies are not confined to fires in forests. Every ecosystem, from tundra to coral reefs, has legacies from past disturbances. Other large-scale and infrequent disturbances such as hurricanes, tornadoes, derechos (downbursts that produce straight-line winds of hurricane force), tsunamis, earthquakes, and volcanoes also leave legacies in whatever ecosystem they travel through. Different types of disturbances in the same ecosystem produce different legacies, and the same disturbance produces different legacies in different ecosystems.[7]

Many of the large legacy-producing disturbances are meteorological or, in the case of fire, are strongly influenced by meteorological events and conditions, such as wind direction or length and severity of drought. These meteorological events, such as hurricanes or tornadoes, are large but still smaller than the grid scale at which global climate models operate. On the other hand, as the earth warms and the energy content of the atmosphere increases, the frequency, severity, locations, and pathways of these disturbance-producing meteorological events will change. Changes in the size, frequency, and locations of these disturbances may be where global climate change meets ecosystems and landscapes most forcefully. Although our global climate models are not yet capable of simulating a hurricane or fire, we are moving in that direction. In the meantime, we can prepare for a time when we can develop a quantitative theory of the coupling of large-scale disturbances and meteorological events by undertaking systematic case studies of legacies in as many ecosystems as we can.

But to do that, we have to learn a new way to look at forests and other ecosystems. When we look at a forest, we usually just see the current trees. To see the legacies the current forest contains, we have to look through it to the ghosts of forests past. The study of how legacies of the past influence the course of the future is only now beginning. What we most need is a sound understanding of how the natural history of fires or other disturbances interacts with the natural history of the landscapes they travel over, the organisms they kill, and the organisms that survive.

21.

Fire Regimes and the Correlated Evolution of Serotiny and Flammability

Some species need fire to disperse their seeds even though the adults are killed. These species also have traits that make them highly flammable. How did this suite of traits evolve? How do these traits determine the fire regime across the landscape?

A number of years ago, I was doing fieldwork in Ottawa National Forest's Sylvania Wilderness Area in Michigan's Upper Peninsula when a few miles east of there, on the sandy outwash near Crystal City, a large fire burned through some forests dominated by jack pine. When I went to see the burn after the fire was out, I saw charred standing snags almost everywhere. Few trees of any species escaped this fire. The landscape could easily be the backdrop for the line from the thirteenth century "Dies Irae" ("Days of Wrath") in the Requiem Mass: "Heaven and Earth in ashes burning."

I went back the next year and saw a landscape that was not a blackened wreck but a green lawn that would have been the envy of Ireland. However, this lawn was composed not of grasses but of millions upon millions of jack pine seedlings. In many places, there were several hundred jack pine seedlings per square meter. Where did these seedlings come from if there were almost no live seed trees left after the fire? The jack pine seed trees certainly were killed by the fire, but it was the fire that opened their cones and enabled their seeds to be released.

Jack pines begin to produce cones when very young, even as young as 5 years old, when they are mere saplings. Their cones curl around the branch to which they are attached and usually remain glued shut and attached for up to several decades. One advantage of closed cones is that the seeds are protected from predators such as crossbills and squirrels. But in order for those seeds to reach the soil and germinate, the cones need to be opened somehow. Some jack pine cones are opened by the heat of the sun, but the cones of most jack pines open only after fire melts the resins, burns the cones, and releases the seeds, which are hidden at the base of the scales. This trait is known as serotiny. Serotiny is not uncommon in pines, especially those closely related to jack pine. Lodgepole pine, a near relative in the Rocky Mountains with which jack pine hybridizes, and pitch pine, which grows along the Atlantic seaboard from Cape Cod to the New Jersey Pine Barrens, also have serotinous cones. Serotinous cones that remain closed until a fire opens them are more common on jack pine in the northern portions of its range, whereas nonserotinous cones that are opened by the sun are more common in the southern portions. Where these two populations overlap, some individuals have both open and closed cones.

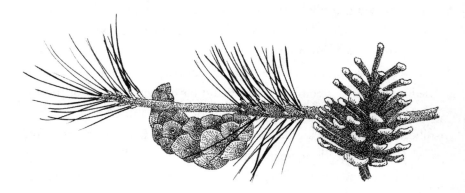

Jack pine has several traits that increase flammability, such as short, resinous needles that burn quickly and the retention of dead lower branches. These two traits enable flames to carry up to the crown. This means that once a fire begins in a jack pine stand, it will almost always burn into the crowns and across the entire stand (see Essay 19). The next year, nearly full sunlight reaches the forest floor, where the seeds have dropped after the fire opened the cones. The seedlings that have subsequently germinated are very intolerant of shade but grow rapidly in full sunlight. Thus, the release of the seeds by the fire, the nearly complete burning of the canopy, and the rapid growth of seedlings in subsequent full sunlight go a long way to ensuring that the next stand will be composed largely of the descendants of the generation of jack pines that burned.

But why aren't the seeds themselves scorched and burned? Why do they remain viable after the cone is charred? The seeds are protected from fire partly because the resins ignite at a low temperature (about 50°C), and the outside of the cone is engulfed with a comparatively cool flame. The interior ends of the scales are also slightly corky, which protects the seeds stored there. The resins are delivered to a pore at the center of the hard outer end of the scale by a duct that extends down through the scale to a small reservoir of resin.[1] The resin burns only when exposed to sufficient oxygen after it exits this pore. The heating of the outer surface of the cone and the relative coolness of the far interior where the seeds are stored form temperature gradients that distort and curl the scales, forcing the cone to blossom open like a flower, which in fact it is.

But the seeds are not shed right away. The melted resin is still sticky while the cone (and the underlying soil surface) is hot, but when the resin cools it shrinks and cracks.[2] At this point the seeds are released onto the now cooled soil, where they can germinate safely. The convergence of these elaborate adaptations—the chemistry of the resins that controls its cool flame when burning and its cracking when cooled, the resin duct to deliver the resin to the surface, and the pore on the outer surface of the scale from which it is exuded—all suggest that natural selection has been at play here.

The convergence of the serotinous cones with the traits that encourage fires to burn once ignited and the rapid growth of millions of seedlings germinating from seeds released by fire certainly seems to be more than coincidental. To explain this convergence of traits, Robert Mutch suggested that serotinous species have evolved them to promote fire and to give the next generation a competitive edge.[3] This hypothesis has been criticized for a number of reasons.[4] First, some of these traits may have been selected for other reasons. For example, resinous needles could have evolved as a defense against herbivores or as protection against scouring by snow and ice. Second, a trait may not make any difference in the transmission of genes to the next generation, in which case it is called selectively neutral. Retention of dead branches may be selectively neutral, for example. Still, Mutch's hypothesis is attractive and cannot be dismissed so easily. A trait need not embody just a single mechanism to enhance reproductive output; in fact, traits that enhance reproductive output by multiple means stand a stronger chance of being selected in subsequent generations.

But perhaps most damagingly, the hypothesis has the whiff of the heresy of group selection about it. Group selection, or the self-sacrifice of some individuals for the perpetuation of the rest of the group or species, including individuals they are not related to, is not widely accepted by the scientific community and is quite controversial. The main argument against group selection is that it makes little sense for an individual to forgo future reproductive output by sacrificing itself. The sacrifice seems to benefit only other individuals that it may not be related to, and so the representation of its genes in the next generation should then decline. But then how can the genes governing traits that lead to self-sacrifice ever spread through the population or species? In the case of jack pine, why should an individual tree forgo future reproductive output by making itself flammable?

There is one case where self-sacrifice does seem to result in a larger representation of an individual's genes in the next generation, and that is if it results not only in more of its own descendants but also descendants of individuals it is related to and therefore shares at least some genes with. This is known as kin selection because the pool of genes that is selected are not only those in an individual but also those in its relatives, or kin. That is, an individual jack pine may appear to sacrifice itself by making its genes go up in smoke, but if more copies of its genes are thereby perpetuated in the seeds released from the burned cones and if those seeds germinate in full sunlight rather than in the shade of adjacent competitors, then the next generation will contain a higher proportion of an individual's genes than if it had lived and produced seeds that died young in the shade of other trees and never bore seeds. William Bond and Jeremy Midgley called this the Kill Thy Neighbor hypothesis.[5]

An individual's fitness includes not only its own genes but also the copies of those genes in its relatives. All trees and most higher plants have two sets of chromosomes that contain the genes. The individual contributes one set of chromosomes to each seed. The other set of chromosomes comes from the individual whose pollen fertilizes the seeds. An individual jack pine therefore shares 50 percent of its genes with its seedlings. So if only three of the seeds released from a burned cone germinate and live to reproductive age, then there will be more copies of the parents' genes in the next generation compared with what there would be if a fire never opened its cones. The many hundreds of seeds released after a fire virtually guarantee that more copies of a parent's genes will be contained in the next generation of seedlings than if a fire had not happened because most of the seeds shed by the few cones that open in the sun's heat would have produced seedlings that would have died in the shade of their parents. The hypothesis of kin selection was first developed by William Hamilton to explain the evolution of apparent altruism in animals, such as the beaver that warns the colony of an approaching predator by slapping its tail on the water but thereby also calls attention to itself.[6] It may be a stretch to think a flammable jack pine commits an act of altruism so that its seeds can be released and get their chance in the sun, but it is thought-provoking nonetheless.

Co-selection of flammability and serotiny implies three testable predictions: First, there needs to be genetically controlled variability in the degree of serotiny between individuals within a population (natural selection can only work when there is variability in a trait). Second, landscapes that are prone to burn should have a greater proportion of serotinous individuals, whereas landscapes not prone to burn should have fewer serotinous individuals. Third, traits that promote flammability should be more common in populations or species with serotinous cones than in populations or species whose cones open without fire.

I mentioned before that there are both serotinous and nonserotinous jack pines as well as some individuals that produce both types of cones. To test whether serotiny is inherited, T. D. Rudolph and colleagues gathered cones from both serotinous and nonserotinous jack pine where the ranges of these two varieties of jack pine overlap in northeastern Minnesota. They then transplanted the germinated seedlings to a common garden where they were all subject to the same soil and weather but were allowed to cross-pollinate freely (see Essay 6, where we also discussed common garden experiments).[7] Fifteen years later, the seedlings had grown into mature trees with several years' worth of cones. Rudolph then tallied the number of progeny with only closed cones, only open cones, and a mix of open and closed cones each parent tree produced. Parents producing only closed and serotinous cones produced five times more progeny with closed cones than with open cones and two and a half times more progeny with closed cones than with mixed cones. Open-coned and nonserotinous parents produced two times more progeny with open cones than with closed or mixed cones. The majority of progeny therefore followed their parent's traits, which is strong evidence that serotiny and nonserotiny are inherited traits.

Further analyses of other populations of jack pine and lodgepole pine by A. H. Teich suggests that the serotinous or nonserotinous traits are governed by two **alleles** of a single gene (Mendel's famous green and yellow pea colors are governed by two alleles of a gene for color, as are blue and brown eyes in humans).[8] Each allele codes for either closed cones or open cones. Trees that are either purely serotinous or nonserotinous have two identical copies of the appropriate allele and so are "pure," or **homozygous**. Trees that have both types of cones have copies of both alleles and are **heterozygous**. Whether or not a cone on these mixed allele trees opens appears to be under the control of the local conditions surrounding each cone, such as whether it is on the sunny or shady side of the tree or whether it is receiving heat reflected off the ground. Although the local environment does have some influence on whether cones open, this study demonstrates that there is strong genetic control over serotiny and that there is strong variability in this trait within populations, allowing selection to occur.

Does fire select for populations of serotinous individuals? Where large numbers of lakes with islands are embedded in the North Woods glacial landscape, there are two different fire regimes: large and lethal crown fires on mainland areas or small and infrequent ground fires on isolated islands. This diversity in the fire regime creates an ideal natural experiment to test whether fire selects for the serotinous and nonserotinous genotypes of jack pine. Sylvia Gauthier and colleagues surveyed twenty-four jack pine populations on islands in Lake Duparquet and the adjacent mainland in western Quebec and indeed found that populations on islands were dominated by nonserotinous individuals, whereas populations on the fire-prone mainland were dominated by serotinous individuals.[9] So fire is a strong selection pressure for or against serotiny in jack pines and its relatives.

Are there correlations between traits that promote flammability and serotiny? Dylan Schwilk and David Ackerly hypothesized that if evolution produced such correlations, then species should be divided into two groups: one in which closed serotinous cones are associated with retention of dead branches that can carry fire into the crown, a short span of years to reproductive maturity so that many closed cones can accumulate in the crown, and needle characteristics that promote burning (such as short needles that dry quickly), and another group that has thick bark that does not burn easily, self-pruning branches so that fire does not carry into the canopy, and nonserotinous open cones so that survivors can disperse seeds for years after a fire.[10] They then analyzed the distribution of these and other characteristics for thirty-eight species in the genus *Pinus* worldwide and found that the species did indeed segregate into these two groups. Serotiny was more common in species with a lower needle density and shorter needles, which should allow air to move more freely throughout the canopy and dry the needles and twigs. Serotiny was also more common in species with a shorter time to reproductive maturity, which would promote the accumulation of many years of closed cones before a fire arrives. Jack pine was in this group. Thicker bark to protect the base of the tree from fire was associated with self-pruning, clear trunks that do not elevate fire into the canopy, higher mature height to keep cones away from the heat of ground fires, and a longer time to reproductive maturity. The nonserotinous and fire-surviving white pine was in this group. Based on associations between these traits, Schwilk and Ackerly were able to construct an evolutionary phylogenetic tree that separated serotinous species such as jack pine with fire-promoting traits from nonserotinous species such as white pine, with fire surviving traits.

So jack pine and other serotinous relatives such as lodgepole and pitch pines pass all three tests of the hypothesis that serotiny is evolutionarily associated with flammability: serotiny has genetic variation that is heritable, fire selects for serotiny, and evolution has selected for associations of closed serotinous cones with traits that promote fire and associations of open nonserotinous cones with traits to survive (but not promote) fires.

It is not hard to imagine serotiny and flammability evolving together, and these studies collectively demonstrate that the fires become more frequent as this suite of traits becomes more common in the population. But fire is part of the abiotic environment, which does not evolve in the usual sense of inheritance of selected genes down through the generations because fire has no genes. Whereas evolution selects for the correlation of the traits that promote fire and the ability to flood the burned area with seeds, the correlation between these traits controls the fire regime in the landscape. The fire regime in turn selects for these traits. Correlated evolution may happen between traits, but this suite of correlated traits may also promote the environmental factors that select for those traits. Evolution may not just select for traits promoting reproductive success along preexisting environmental gradients; it might also produce traits such as serotiny and flammability that promote an environment that selects for them.

This feedback between traits of living organisms and the abiotic properties of the ecosystems they inhabit is an unexplored area of both evolutionary biology and ecosystem ecology. By studying the natural history of trait–environment feedbacks, such as the origin and evolution of serotiny and associated traits, we may be able to merge evolutionary biology with ecosystem ecology.[11] The merger of these two fields by advances in scientific natural history and genetics would be one of the more exciting new developments in ecology.

Climate Change and the Disassembly of the North Woods

Just as the North Woods emerged as the climate warmed after the retreat of the ice sheet, so may climate change cause the disassembly of the North Woods in the next century.

The North Woods often seems eternal to people who live or travel there. This is especially so in the old growth forests of the Boundary Waters Wilderness of northern Minnesota or the Adirondack Park of upstate New York. We stand on the crest of a glacial moraine, look up, and see the old white pines a meter or more in diameter, with century-old fire scars at their bases, seemingly holding up the sky with their massive trunks. We see younger and smaller trees with trunks one-tenth the diameter waiting to take their places in the sun once the old matriarchs die and let some light pour through to the understory. Looking down, we see the dead trunks of previous generations rotting into the forest floor and covered with mosses and mushrooms. Downhill through the pines on the edge of a beaver pond, we may see a cow moose and a calf that look like primeval beasts. We think of bringing our grandchildren to this spot and even our grandchildren bringing their own grandchildren here. In 1954, William Chapman White wrote, in his classic book *Adirondack Country*, "As a man tramples the woods to the lake, he knows he will find pines and lilies, blue heron and golden shiners, shadows on the rocks, and the glint of light on the waters, just as they were in the summer of 1354, as they will be in 2054 and beyond."[1]

Yes, we might see more or less the same scene if we time-traveled back to 1354, but it is an open question whether, in many places, we will be able to see these forests in 2054. In 1354, the carbon dioxide concentration of the atmosphere was a mere 280 parts per million. By 1854, the industrialization of the earth had begun, powered by the burning of coal and oil. This loaded the atmosphere with more carbon dioxide than the earth's plants can take up by photosynthesis. Today, the carbon dioxide concentration of the atmosphere is more than 400 parts per million, and if we keep burning fossil fuels at the rate we are, it will be well above 450 parts per million by 2054. Carbon dioxide traps heat. As its atmospheric concentration is increasing, the temperature of the earth is rising rapidly, especially in northern regions. President George W. Bush said that we are addicted to fossil fuels, and like addicts on some drug, we cannot imagine how we can stop. Like an addict, we are consuming our carbon drugs faster and faster.

As the earth warms, many North Woods species will contract their southern boundaries and expand the northern boundaries of their ranges toward the pole. Spruce and fir probably will invade the regions that are currently tundra. This northward expansion of dark conifers will change the amount of sunlight reflected or absorbed by the earth and its atmosphere. Much of the year, the tundra is now a white expanse of snow and a nearly perfect reflector of the sun's radiation. This white expanse cools the earth. In contrast, the dark conifers to the south of the tundra absorb sunlight and convert it into heat. So as the dark conifers invade the white tundra, more of the sun's warmth will be absorbed and the energy balance of the earth will shift toward warmer falls, winters, and springs.[2] Warming in northern latitudes will not simply produce a shift in these biomes; the shift in species' ranges will also feed back and exacerbate the warming.

In Part I, we learned that the North Woods was not an intact biome sitting south of the ice sheet. Rather, starting 6,000 years ago it gradually assembled itself as the species we know today invaded the barren landscape, one by one and from different directions, as the ice sheet retreated. As we mentioned in Essay 1, the North Woods and its cousin the boreal forest will not simply shift northward but instead are very likely to disassemble with warming as species contract their southern and western boundaries at different rates and in different directions. New species combinations will then replace the North Woods in its current location.

Currently, the southern boundaries of spruce and fir coincide with the southernmost reach of the jet stream, which determines the southern extent of polar air masses in winter.[3] This position of the jet stream is a line from just south of Duluth, Minnesota, through the Upper Peninsula of Michigan, the Adirondacks, and central Vermont, New Hampshire, and Maine. Conversely, the northern boundaries of sugar maple, yellow birch, and other northern deciduous species end where winter temperatures remain below −40°C for prolonged periods. At colder temperatures, the trunks of these species crack, and much of the next summer's carbohydrate production is spent repairing these frost-induced wounds. Temperatures this cold are currently found in northern Minnesota and northern Maine, and sugar maple and yellow birch are much less common north of these latitudes. As the climate warms, the locations of the jet stream, polar air masses, and −40°C temperatures will probably shift and force adjustments in the ranges of these tree species.

Some early models,[4] created in the 1980s, predicted the near disappearance of the North Woods if temperatures increase as little as 2 or 3°C and especially if the warming is accompanied by droughts in midcontinent areas. More recent simulations suggest that suitable habitats for spruce, fir, and jack, red, and white pines could become confined only to the coldest areas of far northern Minnesota; these conifers may exit New England entirely as their southern boundaries retreat north into Canada.[5] Sugar maple and red maple could at first move north into areas vacated by northern conifers. These predictions are already coming to pass, as spruce and fir growth are decreasing and maple growth is increasing in northern Minnesota.[6] But if the temperature increases 4°C or more and prolonged droughts become more common in the midcontinent, sugar maple could also retreat eastward and become confined largely to maritime climates in northern Maine, with perhaps a fringe population remaining along the shore of Lake Superior.

Even in those places, sugar maple might survive only on clay-rich moraines, which can hold sufficient water to support their growth,[7] just as we saw that the pattern of invasion of northern species into the northern landscape after deglaciation depended on the distribution of sandy outwash and clay-rich moraines.[8] If we continue to warm the climate as we have been doing, the different migrations of species may cause the North Woods to become confined to isolated pockets in Canada sometime during my grandson's lifetime, and certainly during my great-grandchildren's lifetimes.

Exactly how all this will unfold is an interesting scientific question. Many of the changes in a species' distribution may be driven by how warming drives the species' phenology, which is the seasonal timing of events in its life cycle. Phenology is a classic branch of natural history that has for too long been considered, as James Watson once said to E. O. Wilson, "mere stamp collecting"[9] but which, as we noted in the Introduction, is becoming increasingly important for how species will respond to climate change. As the spring weather comes earlier and earlier, the timing of some bud breaks, flowerings, emergences from hibernation, spring migration arrivals, and mating behaviors are beginning earlier, by as much as a week or more. As a result, long-term natural history records documenting changing phenologies are assuming increasing importance because these records are concrete evidence that plants and animals are already responding to climate change.[10] Franz Badeck and colleagues call long-term records of phenology, which can be found in many naturalists' field notebooks, "treasures to be discovered."[11] The premier natural history notebooks are, of course, those kept by Thoreau in the forests around Walden Pond. By comparing flowering times today around Walden Pond with those noted by Thoreau in his field notebooks, Robert Primack has learned that understory North Woods plants, such as bunchberry and blueberry, are flowering at least a week and even 2 or 3 weeks earlier now than during Thoreau's time.[12]

Keeping field notes of the changing phenologies of plants and animals from year to year is an excellent way for citizen-scientists or school programs to compile valuable records of responses of local populations to climate change. All it takes are weatherproof field notebooks, field guides for species identification, and a calendar. Observers can now submit data online to the National Phenology Network to be entered into a nationwide database, where they can learn how their observations compare with others.[13]

Besides being cued to climatic signals such as rising or falling temperatures, phenological events such as flowering, unfolding of leaves, emergence from hibernation, and spring can also be cued to astronomical events such as increasing daylength. Although global warming is changing the timing of climate events such as date of last frost, it is not going to change astronomical properties such as daylength. Unfortunately, we still don't understand which of these climatic and astronomical cues are important for most species' phenological development. Some species may need both types of cues to initiate development. Without this basic natural history knowledge, it is difficult to predict how global warming will affect each species except in broad terms such as the shifts in their ranges.

It is even more difficult to predict how the interactions between plants and pollinators, between herbivores and plants, and between predators and prey will be changed if two or more interacting species respond to different cues. The interactions between species determine the roles they play in ecosystems; those interactions depend on how natural selection has synchronized their seasonal developments and life cycles. This synchronization is especially critical in the spring, when insects emerge from hibernation and birds arrive after a long migration and both are desperately in need of food. But if one species of an interacting pair responds to climatic cues while the other responds to daylength, then their life cycles become desynchronized as climate warms but daylength remains the same. The roles these species play in ecosystems then may be lost. We very much need a way to measure the decoupling of the phenologies of interacting species in order to understand and predict the consequences for their coevolution and continued existence. Marcel Visser and Christiaan Both suggest that changes in the timing of the abundance of a species' food may provide such a yardstick.[14]

As an example, the emergence of spruce budworm caterpillars and the young shoots of spruce and fir on which they feed are both largely controlled by temperature, and both are happening earlier with warmer springs.[15] As we saw in Part III, MacArthur's warblers are major predators on the spruce budworm, and their predation helps keep budworm populations in check in most years. In contrast to the emergence of spruce budworm caterpillars, spring migration and arrival of these warblers on their breeding grounds is probably controlled largely by daylength[16] and only weakly by temperature. Despite earlier springs in the north, warblers are not arriving sooner to take advantage of the earlier abundances of caterpillars.[17] Warming has therefore desynchronized the emergence of caterpillars and the spring arrival dates of the warblers, resulting in as many as 20 fewer days for the warblers to feed on the caterpillars before they pupate and metamorphose into adults. In many years, the warblers probably now miss the peak abundance of caterpillars.[18] The control the warblers exert on spruce budworm populations will probably be weakened as a consequence.

As another example, one of the very first butterflies to emerge in spring in northern Minnesota is the mourning cloak. Adults from the previous summer hibernate over the winter in hollow logs or beneath loose pieces of bark on a trunk. Usually, the adults emerge from their winter quarters and mate sometime in April. Females then lay their eggs in rings encircling willow twigs. The caterpillars soon hatch and feast communally on the willow leaves, which begin to unfold a week or two after the eggs are laid. In the 1990s, the average date[19] of emergence of mourning cloaks was April 14, with March 28 or 29 the earliest date of emergence. But in this century, warmer springs have caused mourning cloaks to emerge from hibernation 10 days earlier, and in 2 years (2000 and 2012) as early as March 7. April 4, and certainly March 7, is much earlier than when willow leaves usually unfold. The longer time between earlier emergence of mourning cloaks and the development of their caterpillars and the unfolding of willow leaves may put the caterpillars at increased risk of starvation. The desynchronization of the emergence of hibernating mourning cloaks and the unfolding of willow leaves may spell the demise of local populations of these butterflies.

These are only two examples of how a sounder understanding of the natural history of each species will help us predict how climate change will affect the North Woods and other biomes. But as interesting as these problems and hypotheses are, no scientist whom I know wants the climate to change so that he or she could test them. No one wants to learn how any biome about which we know so little (which is to say all biomes on Earth) is disassembled. No one wants a species to go extinct to see what would happen next. No one has ever spent a lifetime studying a species or the interactions between two or more species only to conclude that the species really are not essential and can safely be disposed of. The more we learn about the natural history of organisms in their landscapes, the more we realize that every species does something essential that cannot be replaced, something that is not completely redundant with other species, something that will be lost forever if that species goes extinct. Life goes on in the face of extinction, but there is a difference between extinctions that have happened in the geological past and extinctions that we knowingly cause but refuse to take responsibility for. The first is part of the natural history of life on this planet; the second has serious moral implications for how we interact with the rest of life on Earth.

So much of how we define ourselves as a people depends on the natural history of the landscapes we live in and the organisms we live with. Arizonans are the people of the Sonoran Desert; Minnesotans are the people of big pines, wolves, moose, and the wails of loons; Vermonters are the people of sugar maples and maple syrup. Who will we be if we lose the landscapes and organisms that define us? What will our grandchildren and great-grandchildren think of us when they learn that, by burning fossil fuels, we deprived them of the opportunity to also be the people of white pine, moose, and loons, even though we knew what the consequences would be? Can any of us look our grandchildren and great-grandchildren in the eye and try to explain this without shame? Renewing the teaching of the natural history of where we live to our children and grandchildren could help turn things around. If our children and grandchildren knew they could be losing the North Woods and many other biomes, they may do a better job of saving them.

The Natural History of Beauty

Keen observations of natural history and our sense of the beautiful arose early in our evolution and are core to our sense of what it means to be a human being.

About 31,000 years ago, in the limestone region of the Ardèche Valley in southern France, people entered a cave—now called Chauvet, after one of its discoverers—and painted images of animals on the walls.[1] These people were exceptional observers of the natural history of these animals. The articulations of the hind legs of reindeer, bison, horses, the now-extinct aurochs, and many other mammals are accurately rendered in these paintings. Many students in my biological illustration class get the articulation of hind legs wrong and have a hard time understanding it until I show them the correspondence between the bones in an ungulate's leg and in a human leg. My students usually think that the joint we see in the middle of a horse's hind leg or that of any other ungulate is a knee, but the leg is bending the wrong way there for it to be a knee. This joint is the ankle (the knee is much higher, near the haunch). The lower leg below this joint is an elongated foot, and the animal walks on its toes. The cave painters always got this correct. This accuracy implies that they understood the anatomy of the animals, perhaps by

246 OF WHAT SHOULD A CLEVER MOOSE EAT?

butchering them, as shown by marks of knife blades on the animal's bones. Moreover, the articulation of the painted animals changes correctly with different gaits. The heads of a herd of horses in one painting at Chauvet are set at the exact angle of horses in gallop, and the mouths are open as if the horses are breathing hard. The people who made these paintings knew, in great detail, the world in which they lived.

These are some of the oldest known paintings anywhere. The descendants of the people who made them followed the retreat of the ice sheets north and wandered over northern Europe, across Asia, and into North America. For many thousands of years, a vast continuum of very similar human cultures spanned the north from Scandinavia, through Siberia, across the Bering Land Bridge and into Alaska, down the Pacific Coast, and eastward along the edge of the Ice Sheet to the Western Great Lakes region, arriving there as the North Woods was beginning to assemble itself. This was the largest and most durable set of cultures of all human history.[2] Every one of these related cultures painted images of the important animals in their worlds in caves and on cliffs. Moose, reindeer, caribou, and bears were common subjects from northern Scandinavia to northern Minnesota. These images are remarkably similar to the paintings in the Chauvet cave in style, workmanship, and the red and yellow ochre minerals used in the paints. In almost every case, the species of animal, and its behavior, is easily recognizable.

Although some of these paintings are crude, as if made by beginners, many are strikingly beautiful. Lines are clean and spare, almost like Japanese brush paintings. Depth is depicted by partially drawn profiles of one animal behind another. The irregularities and stains on the rock surfaces often were used effectively to enhance haunches, the bulging bellies of pregnant females, or the shoulders of bulls in the prime of rut. Not only did these people know the natural history of the animals they painted, they also pushed the limits of their art in the mixing of ochre pigments with fats to make durable paints, in the accuracy and elegance of the composition of their paintings, and in their apparent ability to learn and improve. The best of these paintings convey a sense of awe. When I have taken visitors to see the ancient rock paintings of moose on Hegman Lake in northern Minnesota, their reaction is first a soft "Wow" and then stunned silence.

Keen observations of natural history, wondering about the world around us, and depicting our wonder in works of art arose together early in our evolution, as the Chauvet paintings so eloquently attest. This era is when we became human. The ability to observe and wonder about nature and transfer that to works of art are some of the attributes that define what it means to be a human being today, along with a social structure based on kinship, language, music, and song, among others. We share 95 percent of our genes with chimpanzees, mainly genes for metabolism and development. But chimpanzees do not paint lions, elephants, and other animals on rock walls, and they do not write treatises on the natural history of their environment. Natural history and art, as well as other uniquely human attributes, emerge at least partly from the 5 percent of our genes we don't share with chimpanzees.

A number of naturalists have commented on this confluence of art and natural history.[3] E. O. Wilson claims we have a biological need to love nature and especially to find nature beautiful, which he calls biophilia. It is easy to understand why a detailed knowledge of nature helps people thrive: If you know where and when to find berries or honey, how animals travel, where to thrust a spear into an animal to inflict rapid death, and how to distinguish which plants are useful and which should be avoided, you have a better chance of surviving and caring for your children. The genes that help you obtain and retain this knowledge of the natural history of your environment will then be propagated into future generations.

But what could be the adaptive significance of beauty? We think a work of art is beautiful if its symmetry or asymmetry pleases us, if the juxtaposition of shapes and colors arrests our eyes and surprises us, and if our eyes are led harmoniously through the paintings or across the surfaces of the sculpture. The ability to recognize beautiful things has adaptive significance not because such things are safe—the view from a cliff is beautiful, but you'd better watch your step—but because we are drawn to linger over them and, in doing so, observe them more completely and then later ponder them and invest them with meaning. The Chauvet cave paintings were early humans' attempts to ponder the natural world and understand it by abstracting the forms and behavior of animals into images.

There is a sense of playfulness to these paintings, which were probably made by young people, most likely young men, as Dale Guthrie concluded from analyses of the sizes and shapes of handprints on the cave walls.[4] Play, which is usually done in a safe environment, is a highly stylized version of more serious adult behaviors. As Guthrie notes, play is a safe way to practice dealing with more dangerous situations, such as a hunt for the bison or mammoth. Quite a number of these paintings have lines, probably depicting a spear, thrust acutely into the animal in the exact spot for a mortal wound. It is as if the painter were practicing how he would make the fatal thrust or perhaps reviewing what went right on a recent hunt so that he could repeat it next time. When children play make-believe, they are trying out new combinations of behaviors and objects to see whether they work harmoniously. In the paintings, animals are often overlaid on one another in new combinations as if the painters were trying to imagine connections between them. Many animals in the paintings appear to be leaping into the air or are painted on ceilings as if they were flying in a make-believe sky. In their paintings, the Chauvet people tried to go beyond the superficial observations of their senses, to explore connections between their own cognitive, social, and spiritual worlds and those of the animals. All these features are attributes of play. Play is an abstraction of the world, reducing it to basic essentials.

Does this sound like the practice of science to you? It does to me. The best science and the best art have a sense of play about them. Go into the lab or studio of anyone who is doing good science or good art. The scientist or artist and perhaps their students are playing with new ideas, methods and techniques, brushes and paper, and equipment in a safe but constructively critical atmosphere. Both hypotheses and the rules of composition are abstractions of the world around us, but the best scientists don't just test hypotheses with statistics, and the best artists don't simply apply the rules of composition blindly; they play with the hypotheses and the rules of composition by combining different ideas in new ways. If something doesn't work, they try something else. Scientists and artists alike are looking for abstract patterns that unify the chaos of the world around us and hoping that these patterns are elegant and hence beautiful.

The attributes of beauty in works of art—symmetry or broken symmetry, coherence, harmony, and surprise—are the same attributes of organisms and landscapes that catch our eye and compel us to investigate their natural history more deeply. When we investigate the natural history of an organism or landscape, we uncover new connections between that organism or landscape and other organisms or processes. These connections have their own symmetries, such as the symbiosis in lichens in which algae provide carbohydrates and fungi provide nutrients for their mutual benefit; asymmetries such as spruce trees growing faster than sugar maples on infertile soils; harmonies such as the stability of the lynx–hare cycle; and surprises, of which we have seen many in this book. In short, the deeper investigation of the natural history of organisms and landscapes has all the attributes of beauty.

Although art and science share these attributes and perhaps share an origin in cave paintings, they are not the same thing. The purpose of art is to express the artist's emotional response to the world (by *emotion* I also mean the feelings that accompany intense curiosity). The purpose of science is to construct an objective description and explanation of the natural world about which we can all agree. Objectivity does not forbid a scientist to hold aesthetic feelings about the materials he or she is working with. Instead, objectivity requires that "the effort to see the object as in itself it really is be well and truly made."[5] A scientist's moral courage lies in this intellectual effort to be objective. This is a kind of respect to the small piece of nature we happen to be studying. To show this respect, we must always consider the possibility that our descriptions and theories of the natural world may be wrong. This is what the null hypothesis is all about, and it is why offering a clear null hypothesis and then accepting it when it is correct is an act of moral courage.[6] Art can be aesthetically pleasing or not, it can move us or not, but it cannot be "wrong"; what would "wrong" art look or sound like?

In creating the drawings for this book, I have had to make scientific decisions about which details of natural history to include and artistic decisions about how to include those details in an aesthetically pleasing way. I want the drawings to embody the words in the text so that I can teach you, the reader, something about natural history. But no one wants to look at an ugly work of art. If the drawing is not aesthetically pleasing, I miss the chance to use it to investigate with you some detail of natural history and how that both deepens and expands our view of nature.

Artists often say that science takes away beauty. I have never understood this. Surely, a flower is beautiful or has a beautiful smell to a scientist as well as to an artist. I think much of this attitude may be best expressed in Tennyson's line, "Science grows and beauty dwindles," as if science takes away rather than enriches our perception of natural beauty. I prefer instead the last and equally poetic paragraph of the *Origin of Species*, which begins with the line "There is grandeur in this view of life." That view of life was, of course, natural history.

Clear natural history descriptions serve science by keeping our theories true to nature, but they also serve art by uncovering deeper layers of beauty in the natural world. These layers are the relationships that themselves produce beautiful patterns, and they include those between the parts of organisms, such as the shapes of leaves and their arrangements in canopies, or between species, such as the dance of hare and lynx, or between a species and its physical environment, such as where a beaver builds a dam to create a pond.

There are some wild plums behind our house that have spectacular sprays of snow white flowers in spring. In the evening, they attract many dozens of bees from our hive and other insect pollinators. The pollen from each of these flowers is borne at the end of long, thin stamens that brush the pollen across the entire body of visiting insects, which the insects then transfer to the next flower they visit and cross-pollinate. I sometimes stand beneath these trees and listen to the hum of insects just above my head. The bees are so intent on gathering pollen and nectar that they pay me little attention, so I don't worry about being stung. All this is unmistakably pleasing to me.

The insects are continuously arriving from all directions. As I stand

there watching the cloud of insects encompass the canopy of flowers over my head, I begin to wonder how they find the flowers from far away. The flowers emit perfumes that I can smell, and these scents must be attractive to the insects as they are to me, but am I and the insects smelling the same perfumes? The perfumes are volatile compounds with low molecular weights, which I detect with my olfactory nerves and which bees and insects detect with their antennae. These molecules have simple symmetric shapes, such as five or six carbon rings, but their particular smell to us comes by side chains that break these symmetries. Shift the side chain to a different carbon, and the smell will often change radically to me, as will the behavioral reactions of bees and insects to them.[7] Do the nerves in the antennae of bees emit the same patterns of electrical signals as my olfactory nerve does when exposed to these different molecular structures? Perhaps the insects are also attracted by the striking patterns of the flowers against the dark forest behind them, but they may not see the flowers as we do. Many insects see in ultraviolet light rather than visible light, and if you illuminate a flower in ultraviolet light you may find a different pattern of pigments that is normally invisible to us, but it may be the patterns the insects see. The insects and I are both attracted by the showy flowers and scents of the plums, but we may not be seeing or smelling the same things. If the insects see or smell the several species of wild plum in northern Minnesota differently, that may help explain why these plums rarely hybridize, although they often grow together and look nearly identical to me. From my own observations of bees in our garden, I know that they gather nectar and pollen from only one species of plant at a time. Do the bees perceive these species of plum differently and gather nectar and pollen from only one species at a time and thereby not cross-pollinate them? No one knows the answers to these questions. Here are the beginnings of a research program on the natural history of wild plums and their pollinators.

Go into the woods, meadows, or beaches near your home, look

around, and pay attention. It matters not whether you begin with a datasheet, a field notebook, or a sketchpad. You will be surrounded by organisms that pose questions about their connections to each other and to the landscape. Their natural histories will be equally beautiful and just as pleasing as the wild plums behind our house and the bees from our hive. You will be off and running to a deeper and richer appreciation of the natural history of each organism and its beauty.

Notes

Prologue: The Beauty of Natural History

1. Ghiselin (1969), Stott (2013)
2. Tschinkel and Wilson (2014)
3. Greene (2005)
4. Burkhardt (2008), Burkhardt et al. (2008)
5. Darwin and Costa (2009), pp. 73–74
6. Bonner (1993)
7. Bartholomew (1986)
8. Horn (1971)
9. Judson (1980)
10. Noss (1996), Pyle (2007), Tewksbury et al. (2014)
11. Nabhan and Trimble (1994)
12. Dayton (2003), Tewksbury et al. (2014)
13. Fleischner (2005)
14. Tewksbury et al. (2014)
15. Guthrie (2005)
16. Pyle (2007)

Introduction: The Nature of the North Woods

1. Miller-Rushing and Primack (2008)
2. Pastor and Mladenoff (1992)
3. Lopez (1986)
4. Likens et al. (1977), Bormann and Likens (1979)
5. Kalm (1770)
6. Thoreau (1950)
7. Marsh (1864)
8. Nash (1967)
9. Heinselman (1996)
10. Leopold (1941)

1. Setting the Stage

1. In his poem "Fire and Ice," from Frost (1923)
2. Hutchinson (1965)
3. Agassiz (1840), Imbrie and Imbrie (1986)
4. Agassiz (1875)
5. Thorson (2014)
6. Flint (1971), Dawson (1991)

7. Imbrie and Imbrie (1986)
8. Benn and Evans (2010)
9. Sella et al. (2007)
10. Wu and Peltier (1984)
11. Seppè et al. (2012)
12. Embleton and King (1968), Benn and Evans (2010)
13. Ojakangas (2009)
14. Thorson (2014)
15. Post et al. (1982)
16. Lindeman (1942)

2. The Emergence of the North Woods

1. My palynological friends will cringe, but for keys to identification and clarity and beauty of drawings of pollen grains, I think the best book is still Wodehouse (1935).
2. Davis (1981)
3. Ritchie and MacDonald (1986)
4. Kutzbach and Guetter (1986)
5. Davis (1983)
6. Livingston (1905), Veatch (1928), Curtis (1959), Hanson and Hole (1967), Pastor et al. (1982)
7. Brubaker (1975)
8. Imbrie and Imbrie (1986)
9. Graumlich and Davis (1993)
10. Solomon and Webb (1985), Pastor and Post (1986)
11. Shugart et al. (1981)
12. Swain (1973), Whitney (1986)
13. http://www.neotomadb.org/
14. http://www3.nd.edu/~paleolab/paleonproject/
15. http://contemplativemammoth.wordpress.com/

16. MacDonald et al. (1993)
17. Davis and Shafer (2006)
18. Webb et al. (1983), Davis (1983)
19. Davis (1986)
20. Wright (1984)
21. Jacobson and Bradshaw (1981)
22. Webb (1986)
23. Sturm et al. (2005), Bonan et al. (1992)

3. Beaver Ponds and the Flow of Water in Northern Landscapes

1. Johnston and Naiman (1990)
2. Doucet et al. (1994)
3. Woo and Waddington (1990)
4. Naiman et al. (1988)
5. Pastor et al. (1996)
6. Naiman et al. (1993), Westbrook et al. (2011)
7. Ruedemann and Schoonmaker (1938)

4. David Thompson, the Fur Trade, and the Discovery of the Natural History of the North Woods

1. Newman (1985)
2. Thompson and Moreau (2009)
3. Ibid., p. 25
4. Ibid., p. 84
5. Ibid., p. 191
6. Ibid., p. 160
7. Ibid., p. 84
8. Ibid., p. 114
9. Ibid., footnote 2, p. 114
10. Hundertmark et al. (2002)
11. Thompson and Moreau (2009), p. 191

5. How Long Should a Leaf Live?

1. Chabot and Hicks (1982)
2. Kikuzawa (1991)
3. Portis and Parry (2007)
4. Vitousek and Howarth (1991)
5. Mitchell and Chandler (1939), Pastor et al. (1984)
6. Monk (1966)
7. Niklas (1991), Smith and Brewer (1994)
8. Wright et al. (2004)
9. Funk and Cornwell (2013)
10. Kikuzawa and Lechowicz (2006)

6. The Shapes of Leaves

1. Preston (1948)
2. Horn (1971)
3. Hickey and Wolfe (1975), Hsü (1983)
4. Bailey and Sinnott (1915) seem to be the first of many authors to calculate paleotemperatures from the proportion of toothed leaves.
5. Baker-Brosh and Peet (1997)
6. Royer and Wilf (2006)
7. Wolfe (1993)
8. Royer et al. (2009)
9. Kawamura et al. (2010)
10. Smith and Carter (1998)
11. Sprugel (1989)
12. Royer et al. (2005)

7. The Shapes of Crowns

1. Raup (1942)
2. Cohen and Pastor (1996)
3. Horn (1971)
4. Chazdon and Pearcy (1991)
5. Chen and Black (1992)

6. Evans (1956), Miller and Norman (1971)
7. Aber et al. (1982)
8. Graves and Crawford (1914)
9. Fujita (1928)

8. How Should Leaves Die?

1. Addicott (1982)
2. Harlow (1959)
3. Ryan and Bormann (1982)
4. E.g., Chapin and Kedrowski (1983)
5. Kobe et al. (2005)
6. Killingbeck (1996)
7. McClaugherty et al. (1985), Berg and McClaugherty (2008)
8. Pastor et al. (1984)
9. Whitham et al. (2006)

9. Foraging in a Beaver's Pantry

1. Morgan (1986)
2. Belovsky (1984)
3. McGinley and Whitham (1985), Fryxell and Doucet (1991)
4. Jenkins (1980), Belovsky (1984), Pinkowski (1983)
5. Raffel et al. (2009)
6. Gallant et al. (2004)
7. Johnston and Naiman (1990)
8. Basey et al. (1988)
9. Bailey et al. (2004)

10. Voles, Fungi, Spruce, and Abandoned Beaver Meadows

1. Johnston and Naiman (1990)
2. Sergei Wilde was the first professor of forest soils at the University of Wisconsin. He was retired when

I arrived in the Soils Department as a graduate student, but he remained an active researcher for many years after. "Doc," as everyone called him, was full of interesting ideas, but he could also be a sharp critic of your ideas. Sometimes he was wrong, but you had to think hard about why he was wrong. In that way, he was a very good teacher.

3. Wilde et al. (1950)
4. Trappe and Maser (1976), Maser et al. (1978)
5. Pastor et al. (1996)
6. Terwilliger and Pastor (1999)
7. Gunderson (1959)
8. Clough (1964), Grant (1969), Morris (1969)

11. What Should a Clever Moose Eat?

1. Owen-Smith and Novellie (1982)
2. Peterson (1955)
3. Geist (1974)
4. Krefting (1974)
5. McInnes et al. (1992)
6. Pastor et al. (1993)
7. McNaughton et al. (1997)
8. Persson et al. (2005)
9. De Jager and Pastor (2008)
10. De Jager et al. (2009)
11. Pastor et al. (1998)
12. Moen et al. (1997)

12. Tent Caterpillars, Aspens, and the Regulation of Food Webs

1. Duncan and Hodson (1958)
2. Fitzgerald and Webster (1993)
3. Stevens and Lindroth (2005)
4. Cornell and Hawkins (2003)
5. Tilman (1978)
6. Pulice and Packer (2008)
7. Doak et al. (2007), Young et al. (2010)
8. Mattson and Addy (1975)

13. Predatory Warblers and the Control of Spruce Budworm in Conifer Canopies

1. Ehrlich et al. (1988), Wiens (1989)
2. MacArthur (1958)
3. George and Mitchell (1948)
4. Mitchell (1952)
5. Holling (1978, 1988)
6. Wellington et al. (1950)
7. Fleming and Volney (1995)
8. Ibid.
9. Mattson et al. (1991)
10. Simard and Payette (2001)
11. Baskerville (1975)
12. Morin et al. (1993), Morin (1994)
13. Holling (1978, 1988)
14. Holling (1988)
15. Ibid.

14. The Dance of Hare and Lynx at the Top of the Food Web

1. Huxley (1942)
2. Southwood and Clarke (1991), p. 137
3. Ibid., p. 142
4. Elton (1942)
5. Elton and Nicholson (1942)
6. MacLulich (1937)
7. See Chitty (1948) for a complete

list of all previous annual reports on the Canadian Snowshoe Rabbit Enquiry, and Chitty (1950), which was the last report.
8. Turchin (2003)
9. Stenseth et al. (1997)
10. Krebs et al. (2014)
11. Krebs et al. (1995, 2001)
12. Boonstra et al. (1998)
13. Bryant (1981)
14. Boonstra et al. (1998)
15. Sinclair et al. (1993), Sinclair and Gosline (1997)
16. Crowcroft (1991), p. xii

15. Skunk Cabbages, Blowflies, and the Smells of Spring

1. Heinrich (1993)
2. Knutson (1974), Seymour and Schultze-Motel (1997), Seymour (2004)

16. When Should Flowers Bloom and Fruits Ripen?

1. Rosendahl (1955)
2. Smith (2008)
3. Pollan (2001)
4. Willson and Melampy (1983)
5. Gorchov (1985)
6. Gorchov (1990)

17. Everybody's Favorite Berries

1. Tolvanen and Laine (1995, 1997)

18. Crossbills and Conifer Cones

1. Lack (1947)
2. Grant and Grant (2011)
3. Lack (1944)

4. Benkman (1987a, 1987b, 1993a, 2003)
5. Benkman (2003)
6. Benkman (1993a)
7. Benkman (1993b)
8. Parchman and Benkman (2002)
9. Benkman et al. (2001)
10. Benkman (1993b)

19. Does Fire Destroy or Maintain the North Woods?

1. http://earthobservatory.nasa.gov/IOTD/view.php?id=52130&src=ve
2. Dahlkütter et al. (2014)
3. Arno and Allison-Bunnell (2002)
4. Heinselman (1963, 1970)
5. Heinselman (1973, 1981a, 1981b, 1996)
6. Wilderness Act of 1964 (Public Law 88-577)
7. Leopold (1941)
8. See p. 46 of Heinselman (1996) for photos of this procedure and one of the wedges.
9. Digitized images of these maps, including Bud's handwritten notes in the margins, can be found at https://conservancy.umn.edu/handle/11299/168076, which is maintained by the University of Minnesota Libraries. My thanks to Ryan Mattke, in the John R. Borchert Map Library, for showing me these maps and providing the link.
10. You can see a photo of Bud next to this tree on this trip on the back flap of the dust jacket of Heinsel-

man (1996), pointing (I think) to the hole from which the core was taken.

11. Heinselman (1973)
12. Heinselman (1981a), especially table 1
13. Bormann and Likens (1994)
14. Lorimer (1977)
15. Mladenoff and Baker (1999), Keane et al. (2004)
16. Wolter et al. (2012)

20. The Legacies of a Fire

1. Williams (2006)
2. Franklin et al. (2000)
3. The RCMP flight was part of the official survey of the international border through the Quetico–Superior Region. This was the last section of the borders of the lower forty-eight states to have been surveyed. The Treaty of Paris in 1812 set the U.S.–Canada border west of Lake Superior as the main voyageur canoe route until Lake of the Woods, thence the 49th parallel to the Pacific. The journals of David Thompson (see Essay 4) and Alexander MacKenzie had to be consulted and studied first before this part of the border, the true end of the American frontier, could be located.
4. Keeton and Franklin (2005)
5. Niemelä (1997), Johansson et al. (2011)
6. Murphy and Lehnhausen (1998), Nappi et al. (2003)
7. Foster et al. (1998)

21. Fire Regimes and the Correlated Evolution of Serotiny and Flammability

1. Henry (2002)
2. In the far northern edge of their range, jack pines do not need fires to open their cones. Instead, the resin is shrunken and cracked without being previously melted when temperatures get below −40°C, which is the same as −40°F.
3. Mutch (1970)
4. Snyder (1984)
5. Bond and Midgley (1995)
6. Hamilton (1964)
7. Rudolf et al. (1959)
8. Teich (1970)
9. Gauthier et al. (1996)
10. Schwilk and Ackerly (2001)
11. Whitham et al. (2003)

Epilogue: Climate Change and the Disassembly of the North Woods

1. White (1954)
2. Bonan et al. (1992)
3. Bryson (1966)
4. Emanuel et al. (1985), Solomon (1986)
5. Prasad et al. (2007)
6. Fisichelli et al. (2014)
7. Pastor and Post (1988)
8. Brubaker (1975)
9. Wilson (2006)
10. Parmesan (2006)
11. Badeck et al. (2004)
12. Miller-Rushing and Primack (2008), Primack (2014)
13. https://www.usanpn.org/

14. Visser and Both (2005)
15. Lysyk (1989)
16. Breuner and Wingfield (2000)
17. Strode (2003)
18. Ibid.
19. These dates are based on my own observations and dates in Weber (2006)

Postscript: The Natural History of Beauty

1. Clottes (2003)
2. Jarzombeck (2013)
3. Wilson (1984), Skutch (1992), Orians (2014)
4. Guthrie (2005)
5. Trilling (2000)
6. Pastor (2008)
7. Eisner (2005)

Glossary

ablation, zone of   The zone in the lower altitudes of a glacier or lower latitudes of an ice sheet where losses of ice by melting, evaporation from water, sublimation (passing of water molecules from ice directly to water vapor), and calving of chunks of ice from the snout exceed inputs from snow. The glacier or ice sheet is maintained in the ablation zone only by flow of ice from the zone of accumulation upstream (upglacier) from it.

accumulation, zone of   The zone in the higher altitudes of a glacier or higher latitudes of an ice sheet where input by snow exceeds losses by melting, evaporation, and sublimation. Ice flows from the zone of accumulation into the zone of ablation downstream (downglacier) from it.

allele   One of several forms of a gene that control the outward appearance of a trait. As an example, blue eyes and brown eyes are controlled by two different alleles of a single gene for eye color.

auxin   A class of "master hormones" produced by plants that control which shoots are allowed to grow and which shoots or buds are suppressed as well as the development of other plant parts.

cambium   A sheath of living but unspecialized cells wrapped around the trunk, branches, and twigs of a woody plant that gives rise to bark and phloem cells outward and wood and xylem cells inward from it.

cellulose   A carbon compound consisting of hundreds or thousands of glucose molecules linked into a long chain. Cellulose is a common component of plant cell walls, especially of xylem and phloem cells in wood.

coevolution   The parallel changes in gene frequencies by means of natural selection in two species that mutually interact in predator–prey, pollinator–pollen producer, herbivore–plant, parasite–host, or other ecological interactions.

drumlin   A teardrop-shaped landform consisting of till and formed beneath a glacier or ice sheet. The long tail of the drumlin points in the direction of ice flow. Where many drumlins are associated in proximity to one another, the landscape is called a drumlin field.

epicormic bud   A dormant bud lying beneath the bark of a twig, branch, or stem of a woody plant. The epicormic bud is suppressed by hormones, mainly auxin, produced higher in the shoot or plant, but when the supply of these hormones is suspended (especially after the higher portion of the shoot is removed by a browsing animal), the epicormic bud then sprouts a new side shoot.

epidermis   A layer of cells on the upper and lower surfaces of a leaf, flower, or root that protects the plant tissue from the environment and, in leaves, contains stomata, or pores, which allow the exchange of gases.

erratics   Boulders, cobbles, or stones transported by a glacier or ice sheet to a new location and deposited where the bedrock is of a different rock type than the erratic itself.

esker   A sinusoidal landform with a flat upper surface and steep sides, which was the bed of a meltwater stream that once flowed beneath a glacier or ice sheet.

evolution   Changes in gene frequencies from one generation to the next because of the influences of natural selection, mutation, immigration and emigration, and random survival of individuals acting on the parent generation. Of these four processes, only natural selection results in adaptation and directional change during the evolution of populations or species.

foraging   An animal's search for food in its wild environment.

fractal dimension   A ratio of the change in detail of a pattern to the scale over which a measurement is made. Fractal structures include the branching patterns of trees and shrubs or veins in a leaf.

glaciation   An interval of geologic time, usually thousands of years long, in which glaciers or ice sheets advance. Colloquially known as an ice age.

glycosides   A class of compounds produced by plants in which a sugar molecule is attached to another small molecule that gives it its properties. Glycosides often have toxic effects on animals that eat plants and are thought to be produced by the plant as a defense against these animals.

guild of species   A group of species usually but not always closely related that coexist in the same food web and consume the same resources.

heterozygous   Characteristic of an organism that has two different alleles for a gene. One allele usually is dominant over the other and so is expressed as a trait in the organism; the nonexpressed trait is called recessive.

homozygous   Characteristic of an organism that has two identical alleles for the same gene.

hybrid   An organism that has a combination of alleles from parents of two different varieties or species. In North America, apple, from the genus *Malus*, and crabapple, from the same genus, gave rise to hybrids that are some of the many apple hybrids we eat today.

hybrid swarm   A group of related species whose ranges overlap and which hybridize freely.

Hypsithermal   A warm period of the earth's climate beginning 9,000 years ago and ending 5,000 years ago, whereupon the temperature declined steadily. This subsequent cooling period ended with the current period of global warming, especially since the 1950s. The average global temperature today exceeds the average during the Hypsithermal.

hysteresis   Different pathways of change in an ecosystem depending on whether a controlling factor such as temperature is rising or falling.

ice sheet   A massive glacier that covers a large portion of a continent. At least four ice sheets scoured the landscape where the North Woods lies today, the latest being the Laurentide Ice Sheet. Ice ages are the periods during advances of an ice sheet.

irruption   A rapid spatial extension of a population outside its usual range, often because of a combination of high population density and low food supplies in the core of a species' range. Several bird species have irruptions from the tundra or boreal forest southward into the North Woods during some winters, including crossbills and snowy owls.

kettle lake   A lake that forms when an isolated ice block from a retreating ice sheet melts and the till overlying it collapses into the round depression, or kettle hole. Kettle lakes are often very round instead of irregular.

legacy   A live organism, dead organic debris, or environmental pattern that persists through and after a disturbance such as a fire or hurricane. Legacies provide structure, energy, and seeds for the recovery of the ecosystem from the disturbance.

lignin   A family of large, complex carbohydrate molecules that, along with cellulose, stiffen plant cell walls, especially in woody tissues. Lignin is nutrient poor and difficult to decompose.

mesophyll   The layers of cells internal to a leaf that are sandwiched between the surface epidermal cells. Photosynthesis takes place in mesophyll cells.

moraine   A landform composed of unsorted boulders, cobbles, sand, silt, and clay deposited by a glacier or ice sheet. End moraines are ridges deposited at the snout of an ice sheet; the terminal moraine is the moraine deposited at the furthest advance of the snout. Ground moraines are plastered on the land surface beneath the ice sheet or glacier as it advances. Stagnation moraines are dumped across the landscape during a rapid disintegration of an ice sheet or glacier.

mycorrhizae A symbiotic association between a fungus and a plant root. The plant provides carbohydrates to the fungus, and the fungus takes up additional nutrients shared with the plant.

nectaries Small organs in a flower or sometimes at the base of a leaf petiole that produce nectar and attract insects.

outwash A plain of sandy material deposited by meltwater in front of a glacier or ice sheet.

palynology The study of the long-term history of a terrestrial ecosystem by identification of pollen in layers of sediment in a lake or a peatland.

petiole A stalk attaching the leaf blade to the shoot or stem.

phenology The study of seasonal or annual progression of events in an organism's life cycle.

rumen The first chamber in the digestive system of an animal, most notably in ungulates, where food is first deposited after chewing and fermented. After the rumen is full, the animal finds a place to rest, regurgitates the food, and chews it before swallowing again and depositing the food in the stomach proper for further digestion. The act of regurgitation and the second chewing is called chewing the cud.

serotiny The delayed release of seeds from a protective structure such as a cone after an environmental signal, usually a fire, opens the structure.

snout (of a glacier or ice sheet) The front of the glacier or ice sheet.

speciation The emergence of two or more species from a parent species as natural selection isolates gene exchange between two populations.

stomates Pores in the epidermis of a leaf that allow the exchange of carbon dioxide, water vapor, and other gases between the mesophyll and the atmosphere.

terminus (of a glacier or ice sheet) The location of the snout at any one time.

terpenes A large class of aromatic hydrocarbons, usually with a characteristic and strong resinous smell, such as in turpentine. This class of molecules is thought to deter herbivores from eating leaves and twigs.

till The general term for the unsorted debris of boulders, cobbles, sands, silts, and clays left by a glacier or ice sheet in landforms such as moraines and drumlins.

transpiration The loss of water from the open stomates of leaves. This water is replenished by soil water flowing into roots and being drawn up the xylem, carrying nutrients such as nitrogen, phosphorus, calcium, and magnesium with it.

trophic cascade   The sequence of changes in a food web or the flow of energy and nutrients through it prompted by changes in the species composition or biomass of the topmost (predatory) or bottommost (plant) trophic levels.

trophic level   The position a species occupies in the flow of energy or nutrients in a food web. Species consume organisms or resources at lower trophic levels and are consumed by organisms at higher trophic levels.

ungulate   A large and diverse group of hoofed mammals such as deer, moose, horses, sheep, and goats who are primarily herbivores and digest their food with the aid of a rumen.

xylem   Plant cells that transport water and nutrients dissolved in it from the roots, up through the stem and branches, to the leaves. Wood is composed of dead xylem cells. The walls of xylem cells are composed mostly of cellulose and lignin.

Bibliography

Prologue: The Beauty of Natural History

Bartholomew, G. A. 1986. The role of natural history in contemporary biology. *BioScience* 36: 324–329.

Bonner, J. T. 1993. *Life Cycles: Reflections of an Evolutionary Biologist.* Princeton, NJ: Princeton University Press.

Burkhardt, F., ed. 2008. *Origins: Selected Letters of Charles Darwin, 1822–1859.* Cambridge, England: Cambridge University Press.

Burkhardt, F., S. Evans, and A. Pearn, eds. 2008. *Evolution: Selected Letters of Charles Darwin, 1860–1870.* Cambridge, England: Cambridge University Press.

Darwin, C., and J. Costa, annotator. 2009. *The Annotated Origin of Species.* Cambridge, MA: Harvard University Press.

Dayton, P. K. 2003. The importance of the natural sciences to conservation. *American Naturalist* 162: 1–13.

Fleischner, T. L. 2005. Natural history and the deep roots of resource management. *Natural Resources Journal* 45: 1–13.

Ghiselin, M. T. 1969. *The Triumph of the Darwinian Method.* Oakland: University of California Press.

Greene, H. W. 2005. Organisms in nature as a central focus for biology. *Trends in Ecology and Evolution* 20: 23–27.

Guthrie, R. D. 2005. *The Nature of Paleolithic Art.* Chicago: University of Chicago Press.

Horn, H. S. 1971. *The Adaptive Geometry of Trees.* Princeton Monographs in Population Biology Number 3. Princeton, NJ: Princeton University Press.

Judson, H. F. 1980. *The Search for Solutions.* New York: Holt, Rinehart & Winston.

Nabhan, G. P., and S. Trimble. 1994. *The Geography of Childhood: Why Children Need Wild Places.* Boston: Beacon Press.

Noss, R. F. 1996. The naturalists are dying off. *Conservation Biology* 10: 1–3.

Pyle, R. M. 2007. The rise and fall of natural history. Pages 233–243 in *The Future of Nature*, ed. B. Lopez. Minneapolis, MN: Milkweed Editions.

Stott, R. 2013. *Darwin and the Barnacle.* New York: W.W. Norton and Company.

Tewksbury, J. J., J. G. T. Anderson, J. D. Bakker, T. J. Billo, P. W. Dunwiddie, M. J. Groom, S. E. Hampton, S. S. Herman, D. J. Levey, N. J. Machnicki, C. Martínez del Rio, M. E. Power, K. Rowell, A. K. Salomon, L. Stacey, S. C. Trombulak, and T. A. Wheeler. 2014. Natural history's place in science and society. *BioScience* 64: 300–310.

Tschinkel, W., and E. O. Wilson. 2014. Scientific natural history: telling the epics of nature. *BioScience* 64: 438–443.

Introduction: The Nature of the North Woods

Bormann, F. H., and G. E. Likens. 1979. *Pattern and Process in a Forested Ecosystem.* New York: Springer-Verlag.

Heinselman, M. 1996. *The Boundary Waters Wilderness Ecosystem.* Minneapolis, MN: University of Minnesota Press.

Kalm, P. 1966. *Peter Kalm's Travels in North America.* Translated by A. P. Benson. Reprint. New York: Dover Publications. First published 1770.

Leopold, A. 1991. Wilderness as a land laboratory. Pages 287–289 in *The River of the Mother of God, and Other Essays by Aldo Leopold*, ed. S. L. Flader and J. Baird Callicott. Madison: University of Wisconsin Press. First published 1941.

Likens, G. E., F. H. Bormann, R. S. Pierce, J. S. Eaton, and N. M. Johnson. 1977. *Biogeochemistry of a Forested Ecosystem.* New York: Springer-Verlag.

Lopez, B. 1986. *Arctic Dreams.* New York: Charles Scribner's Sons.

Marsh, G. P. 2003. *Man and Nature.* New York: Charles Scribner's Sons, 1864. Reprint, edited by D. Lowenthal. Seattle: University of Washington Press.

Miller-Rushing, A. J., and R. B. Primack. 2008. Global warming and flowering times in Thoreau's Concord: a community perspective. *Ecology* 89: 332–341.

Nash, R. 1967. *Wilderness and the American Mind.* New Haven, CT: Yale University Press.

Pastor, J., and D. J. Mladenoff. 1992. The southern boreal–northern hardwood forest border. Pages 216–240 in *A Systems Analysis of the Global Boreal Forest*, ed.

H. H. Shugart, R. Leemans, and G. B. Bonan. Cambridge, England: Cambridge University Press.

Thoreau, H. 1950. *The Maine Woods.* Arranged by D. C. Lunt. New York: Bramhall House.

1. Setting the Stage

Agassiz, L. 1840. *Études sur les Glaciers.* Reprinted and translated as *Studies on Glaciers* and published in 1967. New York: Hafner Publishing Company.

Agassiz, L. 1875. *Geological Sketches*, Vol. 2. Boston: Ticknor and Fields.

Benn, D., and D. J. A. Evans. 2010. *Glaciers and Glaciation.* London: Routledge.

Dawson, A. 1991. *Ice Age Earth: Late Quaternary Geology and Climate.* London: Routledge.

Embleton, C., and C. A. M. King. 1968. *Glacial and Periglacial Geomorphology.* London: Edward Arnold.

Flint, R. F. 1971. *Glacial and Quaternary Geology.* New York: John Wiley and Sons.

Frost, R. 1923. *New Hampshire.* New York: Henry Holt & Company.

Hutchinson, G. R. 1965. *The Ecological Theater and the Evolutionary Play.* New Haven, CT: Yale University Press.

Imbrie, J., and K. P. Imbrie. 1986. *Ice Ages: Solving the Mystery.* Cambridge, MA: Harvard University Press.

Lindeman, R. L. 1942. The trophic–dynamic aspect of ecology. *Ecology* 23: 399–418.

Ojakangas, R. W. 2009. *Roadside Geology of Minnesota.* Missoula, MT: Mountain Press Publishing Company.

Post, W. M., W. R. Emanuel, P. J. Zinke, and A. G. Stangenberger. 1982. Soil carbon pools and world life zones. *Nature* 298: 156–159.

Sella, G. F., S. Stein, T. H. Dixon, M. Craymer, T. S. James, S. Mazzotti, and R. K. Dokka. 2007. Observation of glacial isostatic adjustment in "stable" North America with GPS. *Geophysical Research Letters* 34: L02306.

Seppä, H., M. Tikkanen, and J.-P. Mäkiaho. 2012. Tilting of Lake Pielinen, eastern Finland: an example of extreme transgressions and regressions caused by differential post-glacial isostatic uplift. *Estonian Journal of Earth Sciences* 61: 149–161.

Thorson, R. M. 2014. *Walden's Shore: Henry David Thoreau and Nineteenth-Century Science.* Cambridge, MA: Harvard University Press.

Wu, P., and W. R. Peltier. 1984. Pleistocene deglaciation and the earth's rotation: a new analysis. *Geophysical Journal of the Royal Astronomical Society* 76: 753–792.

2. The Emergence of the North Woods

Bonan, G. B., D. L. Pollard, and S. T. Thompson. 1992. Effects of boreal forest vegetation on global climate. *Nature* 359: 716–718.

Brubaker, L. B. 1975. Postglacial forest patterns associated with till and outwash in northcentral Upper Michigan. *Quaternary Research* 5: 499–527.

Curtis, J. T. 1959. *The Vegetation of Wisconsin*. Madison: University of Wisconsin Press.

Davis, M. B. 1981. Quaternary history and the stability of forest communities. Pages 132–153 in *Forest Succession: Concepts and Application*, ed. D. C. West, H. H. Shugart, and D. B. Botkin. New York: Springer-Verlag.

Davis, M. B. 1983. Holocene vegetation history of the Eastern United States. Pages 166–181 in *Late Quaternary Environments of the United States, Volume 2. The Holocene*, ed. H. E. Wright. Minneapolis: University of Minnesota Press.

Davis, M. B. 1986. Climatic instability, time lags, and community disequilibrium. Pages 269–284 in *Community Ecology*, ed. J. Diamond and T. J. Case. New York: Harper and Row Publishers.

Davis, O. K., and D. S. Shafer. 2006. *Sporormiella* fungal spores, a palynological means of detecting herbivore density. *Palaeogeography, Palaeoclimatology, Palaeoecology* 237: 40–50.

Graumlich, L. J., and M. B. Davis. 1993. Holocene variation in spatial scales of vegetation pattern in the Upper Midwest. *Ecology* 74: 826–839.

Hanson, G. F., and F. D. Hole. 1967. *Soil Resources and Forest Ecology of Menominee County, Wisconsin*. Bulletin 85, Soil Series No. 60. Madison: University of Wisconsin Geological and Natural History Survey.

Imbrie, J., and K. P. Imbrie. 1986. *Ice Ages: Solving the Mystery*. Cambridge, MA: Harvard University Press.

Jacobson, G. L., and R. H. W. Bradshaw. 1981. The selection of sites for paleovegetational studies. *Quaternary Research* 16: 80–96.

Kutzbach, J. E., and P. J. Guetter. 1986. The influence of changing orbital parameters and surface boundary conditions on the simulated climate of the past 18,000 years. *Journal of Atmospheric Sciences* 43: 1726–1759.

Livingston, B. E. 1905. The relation of soils to natural vegetation in Roscommon and Crawford Counties, Michigan. *Botanical Gazette* 39: 22–41.

MacDonald, G. M., T. W. D. Edwards, K. A. Moser, R. Pientz, and J. P. Smol. 1993. Rapid response of treeline vegetation and lakes to past climate warming. *Nature* 361: 243–246.

Pastor, J., J. D. Aber, C. A. McClaugherty, and J. Melillo. 1982. Geology, soils, and vegetation of Blackhawk Island, Wisconsin. *American Midland Naturalist* 198: 266–277.

Pastor, J., and W. M. Post. 1986. Influence of climate, soil moisture, and succession on forest carbon and nitrogen cycles. *Biogeochemistry* 2: 3–27.

Ritchie, J. C., and G. M. MacDonald. 1986. The patterns of post-glacial spread of white spruce. *Journal of Biogeography* 13: 527–540.

Shugart, H. H., D. C. West, and W. R. Emanuel. 1981. Patterns and dynamics of forests: an application of simulation models. Pages 74–95 in *Forest Succession: Concepts and Application*, ed. D. C. West, H. H. Shugart, and D. B. Botkin. New York: Springer-Verlag.

Solomon, A. M., and T. Webb III. 1985. Computer-aided reconstruction of late-Quaternary landscape dynamics. *Annual Review of Ecology and Systematics* 16: 63–84.

Sturm, M., T. Douglas, C. Racine, and G. E. Liston. 2005. Changing snow and shrub conditions affect albedo with global implications. *Journal of Geophysical Research* 110: G01004.

Swain, A. M. 1973. A history of fire and vegetation in northeastern Minnesota as recorded in lake sediments. *Quaternary Research* 3: 383–396.

Veatch, J. O. 1928. Reconstruction of forest cover based on soil maps. *Quarterly Bulletin of the Michigan Agricultural Experiment Station* 10: 116–126.

Webb, T. III. 1986. Is vegetation in equilibrium with climate? How to interpret late-Quaternary pollen data. *Vegetatio* 67: 75–91.

Webb, T. III, E. J. Cushing, and H. E. Wright. 1983. Holocene changes in the vegetation of the Midwest. Pages 142–165 in *Late Quaternary Environments of the United States, Volume 2. The Holocene*, ed. H. E. Wright. Minneapolis: University of Minnesota Press.

Whitney, G. G. 1986. Relation of Michigan's presettlement pine forests to substrate and disturbance history. *Ecology* 67: 1548–1559.

Wodehouse, R. P. 1935. *Pollen Grains*. New York: McGraw-Hill.

Wright, H. E. 1984. Sensitivity and response time of natural systems to climate change in the late Quaternary. *Quaternary Science Reviews* 3: 91–131.

3. Beaver Ponds and the Flow of Water in Northern Landscapes

Doucet, C. M., I. T. Adams, and J. M. Fryxell. 1994. Beaver dam and cache composition: Are woody species used differently? *Écoscience* 1: 268–270.

Johnston, C. A., and R. J. Naiman. 1990. Aquatic patch creation in relation to beaver population trends. *Ecology* 71: 1617–1621.

Naiman, R. J., C. A. Johnston, and J. C. Kelley. 1988. Alteration of North American streams by beaver. *BioScience* 38: 753–763.

Naiman, R. J., G. Pinay, C. A. Johnston, and J. Pastor. 1993. Beaver influences on the long-term biogeochemical characteristics of boreal forest drainage networks. *Ecology* 75: 905–921.

Pastor, J., A. Downing, and H. E. Erickson. 1996. Species–area curves and diversity–productivity relationships in beaver meadows of Voyageurs National Park, Minnesota, U.S.A. *Oikos* 77: 399–406.

Ruedemann, R., and W. J. Schoonmaker. 1938. Beaver-dams as geologic agents. *Science* 88: 523–525.

Westbrook, C. J., D. J. Cooper, and B. W. Baker. 2011. Beaver assisted valley formation. *River Research and Applications* 27: 247–256.

Woo, M.-K., and J. M. Waddington. 1990. Effects of beaver dams on subarctic wetland hydrology. *Arctic* 43: 223–230.

4. David Thompson, the Fur Trade, and the Discovery of the Natural History of the North Woods

Hundertmark, K., G. F. Shields, I. G. Udlina, R. T. Bowyer, A. A. Danilkin, and C. C. Schwartz. 2002. Mitochondrial phylogeography of moose (*Alces alces*): Late Pleistocene divergence and population expansion. *Molecular Phylogenetics and Systematics* 22: 375–387.

Newman, P. C. 1985. *Company of Adventurers.* New York: Viking Press.

Thompson, D., and W. E. Moreau, ed. 2009. *The Writings of David Thompson,* Vol. 1. *The Travels, 1850 Version.* Montreal: McGill-Queen's University Press, University of Washington Press, and The Champlain Society.

5. How Long Should a Leaf Live?

Chabot, B. F., and D. J. Hicks. 1982. The ecology of leaf life spans. *Annual Reviews of Ecology and Systematics* 13: 229–259.

Funk, J. L., and W. K. Cornwell. 2013. Leaf traits within communities: context may affect the mapping of traits to function. *Ecology* 94: 1893–1897.

Kikuzawa, K. 1991. A cost–benefit analysis of leaf habit and leaf longevity of trees and their geographical pattern. *The American Naturalist* 138: 1250–1263.

Kikuzawa, K., and M. J. Lechowicz. 2006. Toward synthesis of relationships among leaf longevity, instantaneous photosynthetic rate, lifetime leaf carbon gain, and the gross primary production of forests. *The American Naturalist* 168: 373–383.

Mitchell, H. L., and R. F. Chandler. 1939. The nitrogen nutrition and growth of certain deciduous trees of north-eastern United States. *Black Rock Forest Bulletin* 11.

Monk, C. D. 1966. An ecological significance of evergreenness. *Ecology* 47: 504–505.

Niklas, K. L. 1991. Biomechanical attributes of the leaves of pine species. *Annals of Botany* 68: 253–262.

Pastor, J., J. D. Aber, C. A. McClaugherty, and J. M. Melillo. 1984. Aboveground production and N and P cycling along a nitrogen mineralization gradient on Blackhawk Island, Wisconsin. *Ecology* 65: 256–268.

Portis, A. R., and M. A. J. Parry. 2007. Discoveries in RuBisCO (ribulose-1,5-bisphosphate carboxylase/oxygenase): a historical perspective. *Photosynthesis Research* 94: 121–143.

Smith, W. K., and C. A. Brewer. 1994. The adaptive importance of shoot and crown architecture in conifer trees. *The American Naturalist* 143: 528–532.

Vitousek, P. M., and R. W. Howarth. 1991. Nitrogen limitation on land and in the sea: how can it occur? *Biogeochemistry* 13: 87–115.

Wright, I. J., et al. 2004. The worldwide leaf economics spectrum. *Nature* 428: 821–827.

6. The Shapes of Leaves

Bailey, J. W., and E. W. Sinnott. 1915. A botanical index of Cretaceous and Tertiary climates. *Science* 41: 831–834.

Baker-Brosh, K., and R. K. Peet. 1997. The ecological significance of lobed and toothed leaves in temperate forests trees. *Ecology* 78: 1250–1255.

Hickey, L. J., and J. A. Wolfe. 1975. The bases of angiosperm phylogeny: vegetative morphology. *Annals of the Missouri Botanical Garden* 62: 538–589.

Horn, H. S. 1971. *The Adaptive Geometry of Trees*. Princeton Monographs in Population Biology Number 3. Princeton, NJ: Princeton University Press.

Hsü, J. 1993. Late Cretaceous and Cenozoic vegetation in China, emphasizing their connections with North America. *Annals of the Missouri Botanical Garden* 70: 490–508.

Kawamura, E., G. Horiguchi, and H. Tsukaya. 2010. Mechanisms of leaf tooth formation in *Arapadopsis*. *The Plant Journal* 62: 429–441.

Preston, R. J. Jr. 1948. *North American Trees*. Ames: Iowa State University Press.

Royer, D. L., L. A. Meyerson, K. M. Robinson, and J. M. Adams. 2009. Phenotypic plasticity of leaf shape along a temperature gradient in *Acer rubrum*. *PLOS One* 4: 1–7.

Royer, D. L., and P. Wilf. 2006. Why do toothed leaves correlate with cold climates? Gas exchange at leaf margins provides new insights into a classic paleotemperature proxy. *International Journal of Plant Science* 167: 11–18.

Royer, D. L., P. Wilf, D. A. Janesko, E. A. Kowalski, and D. L. Dilcher. 2005. Correlations of climate and plant ecology to leaf size and shape: potential proxies for the fossil record. *American Journal of Botany* 92: 1141–1151.

Smith, W. K., and G. A. Carter. 1988. Shoot structural effects on needle temperature and photosynthesis in conifers. *American Journal of Botany* 75: 496–500.

Sprugel, D. G. 1989. The relationship of evergreenness, crown architecture, and leaf size. *The American Naturalist* 133: 465–479.

Wolfe, J. A. 1993. A method for obtaining climatic parameters from leaf assemblages. *U.S. Geological Survey Bulletin* 2040.

7. The Shapes of Crowns

Aber, J., J. Pastor, and J. Melillo. 1982. Changes in forest canopy structure along a site quality gradient in southern Wisconsin. *American Midland Naturalist* 108: 256–265.

Chazdon, R. L., and R. W. Pearcy. 1991. The importance of sunflecks for forest understory plants. *BioScience* 41: 760–766.

Chen, J. M., and T. A. Black. 1992. Foliage area and architecture of plant canopies from sunfleck size distributions. *Agricultural and Forest Meteorology* 60: 249–266.

Cohen, Y., and J. Pastor. 1996. Interactions among nitrogen, carbon, plant shape, and photosynthesis. *The American Naturalist* 147: 847–865.

Evans, G. C. 1956. An area survey method of investigating the distribution of light intensity in woodlands, with particular reference to sunflecks. *Journal of Ecology* 44: 391–428.

Fujita, J. 1928. Morning woods. *Poetry* 32: 202.

Graves, M., and C. Crawford. 2014. Rediscovering Jun Fujita on Rainy Lake. *Voyageurs National Park Association Summer 2014 Community Update* (Newsletter), Voyageurs National Park Association, Minneapolis, Minnesota.

Horn, H. S. 1971. *The Adaptive Geometry of Trees.* Princeton, NJ: Princeton University Press.

Miller, E. E., and J. M. Norman. 1971. A sunfleck theory for plant canopies. I. Lengths of sunlit segments along a transect. *Agronomy Journal* 63: 735–738.

Raup, H. M. 1942. Trends in the development of geographic botany. *Annals of the Association of American Geographers* 32: 319–354.

8. How Should Leaves Die?

Addicott, F. T. 1982. *Abscission.* Berkeley: University of California Press.

Berg, B., and C. A. McClaugherty. 2008. *Plant Litter,* 2nd ed. New York: Springer-Verlag.

Chapin, F. S., and R. A. Kedrowski. 1983. Seasonal changes in nitrogen and phosphorus fractions and autumn retranslocation in evergreen and deciduous taiga trees. *Ecology* 64: 376–391.

Harlow, W. M. 1959. *Fruit and Twig Key to Trees and Shrubs.* New York: Dover Publications.

Killingbeck, K. T. 1996. Nutrients in senesced leaves: keys to the search for potential resorption and resorption proficiency. *Ecology* 77: 1716–1727.

Kobe, R. K., C. A. Lepczyk, and M. Iyer. 2005. Resorption efficiency decreases with increasing green leaf nutrients in a global data set. *Ecology* 86: 2780–2792.

McClaugherty, C. A., J. Pastor, J. D. Aber, and J. M. Melillo. 1985. Forest litter decomposition in relationship to soil nitrogen dynamics and litter quality. *Ecology* 66: 266–275.

Pastor, J., J. D. Aber, C. A. McClaugherty, and J. M. Melillo. 1984. Aboveground production and N and P cycling along a nitrogen mineralization gradient on Blackhawk Island, Wisconsin. *Ecology* 65: 256–268.

Ryan, D. F., and F. H. Bormann. 1982. Nutrient resorption in northern hardwood forests. *BioScience* 32: 29–32.

Whitham, T. G., J. K. Bailey, J. A. Schweitzer, S. M. Shuster, R. K. Bangert, C. J. LeRoy, E. V. Lonsdorff, G. J. Allan, S. P. DiFazio, B. M. Potts, D. G. Fischer, C. A. Gehring, R. L. Lindroth, J. C. Marks, S. C. Hart, G. M. Wimp, and S. C. Wooley. 2006. A framework for community and ecosystem genetics: from genes to ecosystems. *Nature Reviews Genetics* 7: 510–523.

9. Foraging in a Beaver's Pantry

Bailey, J. K., J. A. Schweitzer, B. J. Rehill, R. Lindroth, G. D. Martinson, and T. G. Whitham. 2004. Beavers as molecular geneticists: a genetic basis to the foraging of an ecosystem engineer. *Ecology* 85: 603–608.

Basey, J. M., S. H. Jenkins, and P. E. Busher. 1988. Optimal central-place foraging by beavers: tree-size selection in relation to defensive chemicals of quaking aspen. *Oecologia* 76: 278–282.

Belovsky, G. E. 1984. Diet optimization by beaver. *American Midland Naturalist* 111: 209–222.

Fryxell, J. M., and C. M. Doucet. 1991. Provisioning time and central-place foraging in beavers. *Canadian Journal of Zoology* 69: 1308–1313.

Gallant, D., C. H. Bérubé, E. Tremblay, and L. Vasseur. 2004. An extensive study of the foraging ecology of beavers (*Castor canadensis*) in relation to habitat quality. *Canadian Journal of Zoology* 82: 922–933.

Jenkins, S. H. 1980. A size–distance relation in food selection by beavers. *Ecology* 61: 740–746.

Johnston, C. A., and R. J. Naiman. 1990. Browse selection by beaver: effects on riparian forest composition. *Canadian Journal of Forest Research* 20: 1036–1043.

McGinley, M. A., and T. G. Whitham. 1985. Central place foraging by beavers (*Castor canadensis*): a test of foraging predictions and the impact of selective feeding on the growth form of cottonwoods (*Populus fremontii*). *Oecologia* 66: 558–562.

Morgan, L. H. 1986. *The American Beaver: A Classic of Natural History and Ecology*. New York: Dover Publications. First published 1868.

Pinkowski, B. 1983. Foraging behavior of beaver (*Castor canadensis*) in North Dakota. *Journal of Mammalogy* 64: 312–314.

Raffel, T. R., N. Smith, C. Cortright, and A. J. Gatz. 2009. Central place foraging by beavers (*Castor canadensis*) in a complex lake habitat. *American Midland Naturalist* 162: 62–73.

10. Voles, Fungi, Spruce, and Abandoned Beaver Meadows

Clough, G. C. 1964. Local distribution of two voles: evidence for interspecific interaction. *Canadian Field Naturalist* 78: 80–89.

Grant, P. R. 1969. Experimental studies of competitive interaction in a two species system. I. *Microtus* and *Clethrionomys* species in enclosures. *Canadian Journal of Zoology* 47: 1059–1082.

Gunderson, H. L. 1959. Red-backed vole habitat studies in central Minnesota. *Journal of Mammalogy* 40: 405–412.

Johnston, C. A., and R. J. Naiman. 1990. Aquatic patch creation in relation to beaver population trends. *Ecology* 71: 1617–1621.

Maser, C., J. M. Trappe, and R. A. Nussbaum. 1978. Fungal–small mammal interrelationships with emphasis on Oregon coniferous forests. *Ecology* 59: 799–809.

Morris, R. D. 1969. Competitive exclusion between *Microtus* and *Clethrionomys* in the aspen parkland of Saskatchewan. *Journal of Mammalogy* 50: 291–301.

Pastor, J., B. Dewey, and D. P. Christian. 1996. Carbon and nutrient mineralization and fungal spore composition of fecal pellets from voles in Minnesota. *Ecography* 19: 52–61.

Terwilliger, J., and J. Pastor. 1999. Small mammals, ectomycorrhizae, and conifer succession in beaver meadows. *Oikos* 85: 83–94.

Trappe, J. M., and C. Maser. 1976. Germination of spores of *Glomus macrocarpus* (Endogonaceae) after passage through a rodent digestive tract. *Mycologia* 68: 433–436.

Wilde, S. A., C. T. Youngberg, and J. H. Hovind. 1950. Changes in the composition of groundwater, soil fertility, and forest growth produced by the construction and removal of beaver dams. *Journal of Wildlife Management* 14: 122–128.

11. What Should a Clever Moose Eat?

De Jager, N., and J. Pastor. 2008. Effects of moose *Alces alces* population density and site productivity on the canopy geometry of birch *Betula pubescens* and *B. pendula* and Scots pine *Pinus sylvestris*. *Wildlife Biology* 14: 251–262.

De Jager, N., J. Pastor, and A. Hodgson. 2009. Scaling the effects of moose browsing on forage distribution, from the geometry of plant canopies to the landscape. *Ecological Monographs* 79: 281–297.

Geist, V. 1974. On the evolution of reproductive potential in moose. *Naturaliste Canadien* 101: 527–537.

Krefting, L. W. 1974. The ecology of the Isle Royale moose. *University of Minnesota Agricultural Experiment Station Technical Bulletin* 297.

McInnes, P. F., R. J. Naiman, J. Pastor, and Y. Cohen. 1992. Effects of moose browsing on vegetation and litter of the boreal forest, Isle Royale, Michigan, U.S.A. *Ecology* 73: 2059–2075.

McNaughton, S. J., F. F. Banyikwa, and M. M. McNaughton. 1997. Promotion of the cycling of diet-enhancing nutrients by African grazers. *Science* 278: 1798–1800.

Moen, R., J. Pastor, and Y. Cohen. 1997. A spatially-explicit model of moose foraging and energetics. *Ecology* 78: 505–521.

Owen-Smith, N., and P. Novellie. 1982. What should a clever ungulate eat? *The American Naturalist* 119: 151–178.

Pastor, J., B. Dewey, R. Moen, M. White, D. Mladenoff, and Y. Cohen. 1998. Spatial patterns in the moose–forest–soil ecosystem on Isle Royale, Michigan, USA. *Ecological Applications* 8: 411–424.

Pastor, J., B. Dewey, R. J. Naiman, P. F. McInnes, and Y. Cohen. 1993. Moose browsing and soil fertility in the boreal forests of Isle Royale National Park. *Ecology* 74: 467–480.

Persson, I.-L., J. Pastor, K. Danell, and R. Bergström. 2005. Impact of moose population density and forest productivity on the production and composition of litter in boreal forests. *Oikos* 108: 297–306.

Peterson, R. L. 1955. *North American Moose.* Toronto: University of Toronto Press.

12. Tent Caterpillars, Aspens, and the Regulation of Ecosystems

Cornell, H. V., and B. A. Hawkins. 2003. Herbivore responses to plant secondary compounds: a test of phytochemical coevolution theory. *The American Naturalist* 161: 507–522.

Doak, P., D. Wagner, and A. Watson. 2007. Variable extrafloral nectary expression and its consequences in quaking aspen. *Canadian Journal of Botany* 85: 1–9.

Duncan, D. P., and A. C. Hodson. 1958. Influence of the forest tent caterpillar upon the aspen forests of Minnesota. *Forest Science* 4: 71–93.

Fitzgerald, T. D., and F. X. Webster. 1993. Identification and behavioral assays of the trail pheromone of the forest tent caterpillar, *Malacosoma disstria* Hubner (Lepidoptera, Lasiocampidae). *Canadian Journal of Zoology* 71: 1511–1515.

Mattson, A. J., and N. D. Addy. 1975. Phytophagous insects as regulators of forest primary production. *Science* 190: 515–522.

Pulice, C. E., and A. A. Packer. 2008. Simulated herbivory induces extrafloral nectary production in *Prunus avium. Functional Ecology* 22: 801–807.

Stevens, M. T., and R. L. Lindroth. 2005. Induced resistance in the indeterminate growth of aspen (*Populus tremuloides*). *Oecologia* 145: 298–306.

Tilman, D. 1978. Cherries, ants, and tent caterpillars: timing of nectar production in relation to susceptibility of caterpillars to ant predation. *Ecology* 59: 686–692.

Young, B., D. Wagner, P. Doak, and T. Clausen. 2010. Induction of phenolic glycosides by quaking aspen (*Populus tremuloides*) leaves in relation to extrafloral nectaries and epidermal leaf mining. *Journal of Chemical Ecology* 36: 369–377.

13. Predatory Warblers and the Control of Spruce Budworm in Conifer Canopies

Baskerville, G. L. 1975. Spruce budworm: super silviculturalist. *Forestry Chronicle* 62: 339–347.

Ehrlich, P. R., D. S. Dobkin, and D. Wheye. 1988. *The Birder's Handbook.* New York: Simon and Schuster.

Fleming, R. A., and W. J. A. Volney. 1995. Effects of climate change on insect defoliator population processes in Canada's boreal forest: some plausible scenarios. *Water, Air, and Soil Pollution* 82: 445–454.

George, J. L., and R. T. Mitchell. 1948. Calculations on the extent of spruce budworm control by insectivorous insects. *Journal of Forestry* 46: 454–455.

Holling, C. S. 1978. The spruce-budworm/forest-management problem. Chapter 11, pp. 143–182 in *Adaptive Environmental Assessment and Management*, ed. C. S. Holling. New York: John Wiley and Sons.

Holling, C. S. 1988. Temperate forest insect outbreaks, tropical deforestation, and migratory birds. *Memoirs of the Entomological Society of Canada* 146: 21–32.

MacArthur, R. 1958. Population ecology of some warblers in northeastern coniferous forests. *Ecology* 39: 599–619.

Mattson, W. J., R. A. Haack, R. K. Lawrence, and S. S. Slocum. 1991. Considering the nutritional ecology of the spruce budworm in management. *Forest Ecology and Management* 39: 183–210.

Mitchell, R. T. 1952. Consumption of spruce budworm by birds in a spruce forest. *Journal of Forestry* 50: 387–389.

Morin, H. 1994. Dynamics of balsam fir forests in relation to spruce budworm outbreaks in the Boreal Zone of Quebec. *Canadian Journal of Forest Research* 24: 730–741.

Morin, H., D. LaPrise, and Y. Bergeron. 1993. Chronology of spruce budworm outbreaks near Lake Duparquet, Abitibi region, Quebec. *Canadian Journal of Forest Research* 23: 1497–1506.

Simard, M., and S. Payette. 2001. Black spruce decline triggered by spruce budworm at the southern limit of lichen woodland in eastern Canada. *Canadian Journal of Forest Research* 31: 2160–2172.

Wellington, W. G., J. J. Fettes, K. B. Turner, and L. M. Belyea. 1950. Physical and biological indicators of the development of outbreaks of the spruce budworm (*Choristoneura fumiferana* (Clemn.)) and the forest tent caterpillar, *Malacosoma disstria* Hbn (Lepidoptera: Tortricidae; Lasiocampidae). *Canadian Journal of Zoology* 30: 114–127.

Wiens, J. A. 1989. *The Ecology of Bird Communities* (2 vols.). Cambridge, England: Cambridge University Press.

14. The Dance of Hare and Lynx at the Top of the Food Web

Boonstra, R., D. Hik, G. R. Singleton, and A. Tinnikov. 1998. The impact of predator-induced stress on the snowshoe hare cycle. *Ecological Monographs* 68: 371–394.

Bryant, J. P. 1981. Phytochemical deterrence of snowshoe hare browsing by adventitious shoots of four Alaskan trees. *Science* 213: 889–890.

Chitty, H. 1948. The snowshoe rabbit enquiry, 1943–46. *Journal of Animal Ecology* 17: 39–44.

Chitty, H. 1950. The snowshoe rabbit enquiry, 1946–48. *Journal of Animal Ecology* 19: 15–20.

Crowcroft, P. 1991. *Elton's Ecologists: A History of the Bureau of Animal Population.* Oxford, England: Oxford University Press.

Elton, C. 1942. *Voles, Mice, and Lemmings.* Oxford, England: Oxford University Press.

Elton, C., and M. Nicholson. 1942. The ten-year cycle in numbers of the lynx in Canada. *Journal of Animal Ecology* 11: 215–244.

Huxley, J. 1942. *Evolution: The Modern Synthesis.* London: Allen and Unwin (reprinted in 2010 by MIT Press).

Krebs, C. J., R. Boonstra, S. Boutin, and A. R. E. Sinclair. 2001. What drives the 10-year cycle of snowshoe hare? *BioScience* 51: 25–35.

Krebs, C. J., S. Boutin, R. Boonstra, A. R. E. Sinclair, J. N. M. Smith, M. R. T. Dale, K. Martin, and R. Turkington. 1995. Impact of food and predation on the snowshoe hare cycle. *Science* 269: 1112–1115.

Krebs, C. J., J. Bryant, K. Kielland, M. O'Donaghue, F. Doyle, S. Carriere, D. SiFolco, N. Berg, R. Boonstra, S. Boutin, A. J. Kenney, D. G. Reid, K. Bodony, J. Putera, H. K. Timm, T. Burke, J. A. K. Maier, and H. Golden. 2014. What factors determine cyclic amplitude in the snowshoe hare (*Lepus americanus*) cycle? *Canadian Journal of Zoology* 92: 1039–1048.

MacLulich, D. A. 1937. *Fluctuations in the Number of the Varying Hare* (Lepus americanus). Toronto: University of Toronto Press.

Sinclair, A. R. E., and J. M. Gosline. 1997. Solar activity and mammal cycles in the northern hemisphere. *The American Naturalist* 149: 776–784.

Sinclair, A. R. E., J. M. Gosline, G. Holdsworth, C. J. Krebs, S. Boutin, J. N. M. Smith, R. Boonstra, and M. Dale. 1993. Can the solar cycle and climate synchronize the snowshoe hare cycle in Canada? Evidence from tree rings and ice cores. *The American Naturalist* 141: 173–198.

Southwood, R., and J. R. Clarke. 1991. Charles Sutherland Elton. 29 March 1900–1 May 1991. *Biographical Memoirs of Fellows of the Royal Society* 45: 130–146.

Stenseth, N. C., W. Falck, O. N. Bjørnstad, and C. J. Krebs. 1997. Population regulation in snowshoe hare and Canada lynx: asymmetric food web configuration between hare and lynx. *Proceedings of the National Academy of Sciences U.S.A.* 94: 5147–5152.

Turchin, P. 2003. *Complex Population Cycles.* Princeton, NJ: Princeton University Press.

15. Skunk Cabbage, Blowflies, and the Smells of Spring

Heinrich, B. 1993. *The Hot-Blooded Insects: Strategies and Mechanisms of Thermoregulation.* Cambridge, MA: Harvard University Press.

Knutson, R. 1974. Heat production and temperature regulation in eastern skunk cabbage. *Science* 186: 746–747.

Seymour, R. E. 2004. Dynamics and precision of thermoregulatory responses of eastern skunk cabbage *Symplocarpus foetidus. Plant, Cell, and Environment* 27: 1014–1022.

Seymour, R. E., and P. Schultze-Motel. 1997. Heat-producing flowers. *Endeavor* 21: 125–129.

16. When Should Flowers Bloom and Fruits Ripen?

Gorchov, D. L. 1985. Fruit ripening asynchrony is related to variable seed number in *Amelanchier* and *Vaccinium. American Journal of Botany* 72: 1939–1943.

Gorchov, D. L. 1990. Pattern, adaptation, and constraint in fruiting synchrony within vertebrate-dispersed woody plants. *Oikos* 58: 169–180.

Pollan, M. 2001. *The Botany of Desire.* New York: Random House.

Rosendahl, C. O. 1955. *Trees and Shrubs of the Upper Midwest.* Minneapolis: University of Minnesota Press.

Smith, W. R. 2008. *Trees and Shrubs of Minnesota.* Minneapolis: University of Minnesota Press.

Willson, M. F., and M. N. Melampy. 1983. The effect of bicolored fruit displays on fruit removal by avian frugivores. *Oikos* 41: 27–31.

17. Everybody's Favorite Berries

Tolvanen, A., and K. Laine. 1995. Aboveground growth habits of two *Vaccinium* species in relation to habitat. *Canadian Journal of Botany* 73: 465–473.

Tolvanen, A., and K. Laine. 1997. Effects of reproduction and artificial herbivory on vegetative growth and resource levels in deciduous and evergreen dwarf shrubs. *Canadian Journal of Botany* 75: 656–666.

18. Crossbills and Conifer Cones

Benkman, C. W. 1987a. Crossbill foraging behavior, bill structure, and patterns of food availability. *Wilson Bulletin* 99: 351–368.

Benkman, C. W. 1987b. Food profitability and the foraging ecology of crossbills. *Ecological Monographs* 57: 251–267.

Benkman, C. W. 1993a. Adaptation to single resources and the evolution of crossbill (*Loxia*) diversity. *Ecological Monographs* 63: 305–325.

Benkman, C. W. 1993b. The evolution, ecology, and decline of the red crossbill of Newfoundland. *American Birds* 47: 225–229.

Benkman, C. W. 2003. Divergent selection drives the adaptive radiation of crossbills. *Evolution* 57: 1176–1181.

Benkman, C. W., W. C. Holiman, and J. W. Smith. 2001. The influence of a competitor on the geographic mosaic of coevolution between crossbills and lodgepole pine. *Evolution* 55: 282–294.

Grant, P. R., and B. R. Grant. 2011. *How Species Multiply: The Radiation of Darwin's Finches*. Princeton, NJ: Princeton University Press.

Lack, D. 1944. Correlation between beak and food in the crossbill, *Loxia curvirostra* Linnaeus. *Ibis* 86: 268–269 and 522–523.

Lack, D. 1947. *Darwin's Finches*. Oxford, England: Oxford University Press.

Parchman, T. L., and C. W. Benkman. 2002. Diversifying coevolution between crossbills and black spruce on Newfoundland. *Evolution* 56: 1663–1672.

19. Does Fire Destroy or Maintain the North Woods?

Arno, S. F., and S. Allison-Bunnell. 2002. *Flames in Our Forest: Disaster or Renewal?* Washington, DC: Island Press.

Bormann, F. H., and G. E. Likens. 1994. *Pattern and Process in a Forested Ecosystem*. New York: Springer-Verlag.

Dahlkötter, F., M. Gysel, D. Sauer, A. Minikin, R. Baumannm, P. Seifert, A. Ansmann, M. Fromm, C. Voigt, and B. Weinzierl. 2014. The Pagami Creek smoke plume after long-range transport to the upper troposphere over Europe: aerosol properties and black carbon mixing state. *Atmospheric Chemistry and Physics* 14: 6111–6137.

Heinselman, M. L. 1963. Forest sites, bog processes, and peatland types in the Glacial Lake Agassiz Region, Minnesota. *Ecological Monographs* 33: 327–374.

Heinselman, M. L. 1970. Landscape evolution, peatland types, and the environment in the Lake Agassiz peatlands. *Ecological Monographs* 40: 235–261.

Heinselman, M. L. 1973. Fire in the virgin forests of the Boundary Waters Canoe Area, Minnesota. *Quaternary Research* 3: 329–382.

Heinselman, M. L. 1981a. Fire and succession in the conifer forests of northern North America. Pages 374–405 in *Forest Succession: Concepts and Applications*, ed.

D. C. West, H. H. Shugart, and D. B. Botkin. New York: Springer-Verlag.

Heinselman, M. L. 1981b. Fire intensity and frequency as factors in the distribution and structure of northern ecosystems. Pages 7–57 in *Fire Regimes and Ecosystem Properties*, ed. H. A. Mooney, T. M. Bonnicksen, N. L. Christensen, J. E. Lotan, and W. A. Reiners. General Technical Report WO-26. Washington, DC: U.S. Department of Agriculture Forest Service.

Heinselman, M. L. 1996. *The Boundary Waters Wilderness Ecosystem.* Minneapolis: University of Minnesota Press.

Keane, R. E., G. J. Cary, I. D. Davis, M. D. Flannigan, R. H. Gardner, S. Lavorel, J. M. Lenihan, C. Li, and T. S. Rupp. 2004. A classification of landscape fire succession models: spatial simulations of fire and vegetation dynamics. *Ecological Modelling* 179: 3–27.

Leopold, A. 1991. Wilderness as a land laboratory. Pages 287–289 in *The River of the Mother of God, and Other Essays by Aldo Leopold*, ed. S. L. Flader and J. Baird Callicott. Madison: University of Wisconsin Press. First published 1941.

Lorimer, C. G. 1977. The presettlement forest and natural disturbance cycle of northeastern Maine. *Ecology* 58: 139–148.

Mladenoff, D. J., and W. L. Baker, eds. 1999. *Spatial Modeling of Forest Landscape Change: Approaches and Applications.* Cambridge, England: Cambridge University Press.

Wolter, P. T., B. R. Sturtevant, B. R. Miranda, S. M. Leitz, P. A. Townsend, and J. Pastor. 2012. *Forest land cover change (1975–2000) in the Greater Border Lakes Region.* Research Map NRS-3. Newtown Square, PA: U.S. Department of Agriculture, Forest Service, Northern Research Station.

20. The Legacies of a Fire

Foster, D. R., D. H. Knight, and J. F. Franklin. 1998. Landscape patterns and legacies resulting from large, infrequent forest disturbances. *Ecosystems* 1: 497–510.

Franklin, J. F., D. Lindenmeyer, J. A. MacMahon, A. McKee, J. Magnuson, D. A. Perry, R. Waide, and D. Foster. 2000. Threads of continuity. *Conservation in Practice* 1: 8–17.

Johansson, T., J. Andersson, J. Hjältén, M. Dynesius, and F. Ecke. 2011. Short-term responses of beetle assemblages to wildfire in a region with more than 100 years of fire suppression. *Insect Conservation and Diversity* 4: 142–151.

Keeton, W. S., and J. F. Franklin. 2005. Do remnant old-growth trees accelerate rates of succession in mature Douglas-fir forests? *Ecological Monographs* 75: 103–118.

Murphy, E. C., and W. Lehnhausen 1998. Density and foraging ecology of woodpeckers following a stand-replacement fire. *Journal of Wildlife Management* 62: 1359–1372.

Nappi, A., P. Drapeau, J.-F. Giroux, and J.-P. L. Savard. 2003. Snag use by foraging black-backed woodpeckers (*Picoides arcticus*) in a recently burned eastern boreal forest. *The Auk* 120: 505–511.

Niemelä, J. 1997. Invertebrates and boreal forest management. *Conservation Biology* 11: 601–610.

Williams, M. 2006. *Making a Poem: Some Thoughts about Poetry and the People Who Write It*. Baton Rouge: Louisiana State University Press.

21. Fire Regimes and the Correlated Evolution of Serotiny and Flammability

Bond, W. J., and J. J. Midgley. 1995. Kill thy neighbor: an individualistic argument for the evolution of flammability. *Oikos* 73: 79–85.

Gauthier, S., Y. Bergeron, and J.-P. Simon. 1996. Effects of fire regime on the serotiny level of jack pine. *Journal of Ecology* 84: 539–548.

Hamilton, W. D. 1964. The genetic evolution of social behavior. I. *Journal of Theoretical Biology* 7: 1–6.

Henry, J. D. 2002. *Canada's Boreal Forest*. Washington, DC: Smithsonian Press.

Mutch, R. W. 1970. Wildland fires and ecosystems: a hypothesis. *Ecology* 51: 1046–1051.

Rudolf, T. D., R. E. Schoenike, and T. Schantz-Hansen. 1959. Results of one-parent progeny tests relating to the inheritance of open and closed cones in jack pine. *Minnesota Forestry Notes* 78.

Schwilk, D. W., and D. D. Ackerly. 2001. Flammability and serotiny as strategies: correlated evolution in pines. *Oikos* 94: 326–336.

Snyder, J. R. 1984. The role of fire: Mutch ado about nothing? *Oikos* 43: 404–405.

Teich, A. H. 1970. Cone serotiny and inbreeding in natural populations of *Pinus bansiana* and *Pinus contorta*. *Canadian Journal of Botany* 48: 1805–1809.

Whitham, T. G., W. P. Young, G. D. Martinsen, C. A. Gehring, J. A. Schweitzer, S. M. Shuster, G. M. Wimp, D. G. Fischer, J. K. Bailey, R. L. Lindroth, S. Woolbright, and C. R. Kuske. 2003. Community and ecosystem genetics: a consequence of the extended phenotype. *Ecology* 84: 559–573.

Epilogue: Climate Change and the Disassembly of the North Woods

Badeck, F.-W., A. Bondeau, K. Böttcher, D. Docktor, W. Lucht, J. Schaber, and S. Sitch. 2004. Responses of spring phenology to climate change. *New Phytologist* 162: 295–309.

Bonan, G. B., D. L. Pollard, and S. T. Thompson. 1992. Effects of boreal forest vegetation on global climate. *Nature* 359: 716–718.

Breuner, C. W., and J. C. Wingfield. 2000. Rapid behavioral response to corticosterone varies with photoperiod and dose. *Hormones and Behavior* 37: 23–30.

Brubaker, L. B. 1975. Postglacial forest patterns associated with till and outwash in northcentral Upper Michigan. *Quaternary Research* 5: 499–527.

Bryson, R. A. 1966. Air masses, streamlines, and the boreal forest. *Geographical Bulletin* 8: 228–269.

Emanuel, W., H. Shugart, and M. P. Stevenson. 1985. Climatic change and the broad-scale distribution of terrestrial ecosystem complexes. *Climatic Change* 7: 29–43.

Fisichelli, N. A., L. E. Frelch, and P. B. Reich. 2014. Temperate tree expansion into adjacent boreal forest patches facilitated by warmer temperatures. *Ecography* 37: 152–161.

Lysyk, T. J. 1989. Stochastic model of eastern spruce budworm (Lepidoptera: Tortricidae) phenology on white spruce and balsam fir. *Journal of Economic Entomology* 84: 1161–1168.

Miller-Rushing, A. J., and R. B. Primack. 2008. Global warming and flowering times in Thoreau's Concord: a community perspective. *Ecology* 89: 332–341.

Parmesan, C. 2006. Ecological and evolutionary responses to recent climate change. *Annual Review of Ecology, Evolution, and Systematics* 37: 637–669.

Pastor, J., and W. M. Post. 1988. Response of northern forests to CO_2-induced climatic change. *Nature* 334: 55–58.

Prasad, A. M., L. R. Iverson, S. Matthews, and M. Peters. 2007–ongoing. *A Climate Change Atlas for 134 Forest Tree Species of the Eastern United States* [database]. http://www.nrs.fs.fed.us/atlas/tree. Delaware, OH: Northern Research Station, USDA Forest Service.

Primack, R. B. 2014. *Walden Warming: Climate Change Comes to Thoreau's Woods.* Chicago: University of Chicago Press.

Solomon, A. M. 1986. Transient response of forests to CO_2-induced climate change: simulation modeling experiments in eastern North America. *Oecologia* 68: 567–579.

Strode, P. K. 2003. Implications of climate change for North American wood warblers (Parulidae). *Global Change Biology* 9: 1137–1144.

Visser, M. E., and C. Both. 2005. Shifts in phenology due to global climate change: the need for a yardstick. *Proceedings of the Royal Society B* 272: 2561–2569.

Weber, L. 2006. *Butterflies of the North Woods: Minnesota, Wisconsin, and Michigan.* Duluth, MN: Kollath-Stensaas Publishing Company.

White, W. C. 1954. *Adirondack Country*. New York: Alfred Knopf.

Wilson, E. O. 2006. *Naturalist*. Washington, DC: Island Press.

Postscript: The Natural History of Beauty

Clottes, J. 2003. *Chauvet Cave: The Art of Earliest Times*. Salt Lake City: University of Utah Press.

Eisner, T. 2005. *For the Love of Insects*. Cambridge, MA: Harvard University Press.

Guthrie, R. D. 2005. *The Nature of Paleolithic Art*. Chicago: University of Chicago Press.

Jarzombeck, M. 2013. *Architecture of First Societies*. New York: John Wiley.

Orians, G. H. 2014. *Snakes, Sunrises, and Shakespeare*. Chicago: University of Chicago Press.

Pastor, J. 2008. The ethical basis of the null hypothesis. *Nature* 453: 1177.

Skutch, A. F. 1992. *Origins of Nature's Beauty*. Austin: University of Texas Press.

Trilling, L. 2000. *The Moral Obligation to Be Intelligent: Selected Essays*. New York: Farrar, Strauss, and Giroux.

Wilson, E. O. 1984. *Biophilia*. Cambridge, MA: Harvard University Press.

Index

f in the index refers to drawings in the text.